Career Opportunities in GEOLOGY and the EARTH SCIENCES

Lisa A. Rossbacher

ARCO PUBLISHING, INC.
NEW YORK

Published by Arco Publishing, Inc.
215 Park Avenue South, New York, N.Y. 10003

Copyright © 1983 by Lisa A. Rossbacher

Library of Congress Cataloging in Publication Data

Rossbacher, Lisa A.
   Career opportunities in geology and the earth
sciences.

   Includes index.
   1. Geology—Vocational guidance.   2. Earth sciences—
Vocational guidance.   I. Title.
QE34.R67   550'.23   82-1799
ISBN 0-668-05205-8 (Reference Text) AACR2
ISBN 0-668-05220-1 (Paper Edition)

Printed in the United States of America

10 9 8 7 6 5 4 3 2 1

Contents

Preface

When I was in college, I had a friend who encouraged me to lie when anyone asked me what I was studying. When I told the truth and said "Geology," the conversation always faltered. A surprising number of people had no clear idea of what geology was, and the usual response was a vague "My, isn't that interesting!"

Today, when I admit I am a geologist, the general conversation usually changes to geology. Is California going to have an earthquake soon? When? What about Mount St. Helens? What causes sinkholes? Where are gold deposits? Is there really an energy crisis? It seems like everyone has at least one question for a geologist.

Geologists no longer "just hammer rocks." Geologists and other earth scientists are discovering oil, prospecting for minerals, and trying to predict earthquakes. A geologist has stood on the surface of the Moon, and earth scientists watched with amazement as the first Viking and Voyager photographs of Mars, Jupiter, and Saturn were returned. Earth scientists are helping find ways to prevent famine, cope with drought, and understand climatic change. Indeed, the quality of life today is directly affected by the work of geologists and earth scientists everywhere. As the quality of life depends more heavily on these scientists, the demand for qualified individuals will continue to increase. It is an exciting, growing field that contributes—both directly and in ways we seldom realize—to society. This book will show you that geology and the other earth sciences are exciting and worthwhile career choices.

This book benefited from information gathered at several meetings, including the 1980 meeting of the Geological Society of America and the 1981 special symposium on employment opportunities and career workshop on Women in Science sponsored by the National Science Foundation and the Aerospace Corporation. Career guidance information from numerous sources was helpful, but Barbara Jefferson of the Career Planning and Placement Office at Whittier College deserves special mention also for directing me to much useful material. The American Geological Institute also deserves special mention for the career information that they make available through the Institute's Education Department and the Women Geoscientists' Committee. The quotations from professional earth scientists were offered cheerfully, and I want to express my gratitude to these men and women as well.

Many individuals helped in the construction of this book. Several of them will be mentioned here, but I am no less grateful to all the others who contributed, often without knowing it. My debt extends back to all the teachers I have had who gave me a feeling of excitement about science, from E. B. Young and R. Hodge to my professors at Dickinson College, the State University of New York at Binghamton, and Princeton University. Others provided more direct contributions. Mason Hill was a source of encouragement and information, as were Alex Baird and his students at Pomona College, who served as attentive guinea pigs as I sorted out my ideas. Bob Mecoy, editor at Arco Publishing, and his policy of positive reinforcement were also helpful. Ed Griffith, Director of the Computer Center at Whittier College, provided invaluable advice and computer time for text editing. Jean and Richard Rossbacher commented constructively on early drafts of the manuscript, and later versions profited greatly from comments by Bill Wadsworth, Mason Hill, and Dallas Rhodes.

I would especially like to thank Whittier College for support in the form of space, time, and encouragement to pursue the research for this book.

Particular gratitude goes to Dallas Rhodes, my husband and colleague, whose humor, interest, repeated readings, and tolerance for strewn chapter fragments helped make this book possible.

CHAPTER 1

The Future in Geology and the Earth Sciences

Civilization exists by geological consent, subject to change without notice.

—Will Durant

Newspaper headlines remind us daily of the current world energy crisis—a shortage of fossil fuels that can only be expected to get worse as time goes by. Rising costs for natural gas, coal, and other energy forms underline this problem. Vying with the energy crisis for attention in the newspapers, television, and radio are a variety of natural disasters: floods, droughts, earthquakes, tornadoes, volcanic eruptions, sinkholes, hurricanes, landslides . . . and the list goes on.

What do these discouraging headlines have in common? Energy and natural hazards are both in the realm of "earth science." This is a vital and exciting scientific field that affects each of us, every day. Earth science includes studies of earth's interior with its geothermal energy sources; its near-surface crust, where all of our undiscovered nonrenewable energy and mineral resources lie hidden; its surface, where we live in contact with the rivers, oceans, mountains, valleys, rocks, soil, and vegetation; and its atmosphere, which provides the air we breathe, produces our weather, and protects us from incoming ultraviolet radiation.

Early humans depended directly upon a few natural resources for food and shelter. As science and technology advanced, modern civilization lost much of its direct contact with and sensitivity to nature. Nature has become an adversary—something for us to

1

control or be protected from. Because of this change in attitude, enhanced by the increasing pressures of an ever-growing global population, many natural processes are now viewed as "natural hazards." Francis Bacon, in 1620, stated this dilemma well when he wrote, "Nature, to be commanded, must be obeyed."

An example of this change in attitude can be found in man's relationship to rivers. Rivers normally overflow their banks on an average of once every year or two. As long as the river's floodplain has not been developed by man, this normal geologic event is allowed to continue undisturbed. Man's earliest civilizations were founded along rivers because the flood plains were fertilized by the silt from each year's flood. When pressures for space result in *construction* on the floodplain, an average—and completely normal—yearly flood becomes a natural hazard, threat to life and property! In the same way, natural events like landsliding, wave action, hurricanes, volcanic eruptions, and the faulting that produces earthquakes have become greater hazards in recent years as we have occupied and developed potentially hazardous areas.

In part because of the increased use of land areas that are only marginally suited for construction, the need for geologists and other earth scientists is at an all-time high. Scientists who can evaluate the risks of various land uses and who can offer knowledgeable advice are needed to prevent natural processes from becoming natural hazards because of poor planning.

Another major need for earth scientists, especially geologists and geophysicists, is related to the growing demand for energy throughout the world. Increasing prices for petroleum on the world market are creating a greater demand for new domestic sources, especially as existing reserves are exhausted. This stimulates exploration for remaining domestic petroleum resources, as well as research in exciting new technologies for the extraction of more oil from known oil fields and unconventional sources. The development of technology for secondary or tertiary oil recovery and production of petroleum from oil shale and tar sands are subjects of intense investigation. Furthermore, we cannot depend on traditional energy sources forever. Oil and natural gas are limited and nonrenewable resources, but we do not yet know what their limits of availability are. While new energy sources are being developed, including geothermal, solar, tidal, and wind energies, we will continue to depend on fossil fuels for energy.

Geologist Harrison Schmitt in Taurus-Littrow Valley,
Apollo XVII Lunar Mission. (Photo courtesy of the
U.S. Geological Survey.)

Geologists and other earth scientists are already playing important
roles in the changing combinations of energy sources that are
being and will be used.

As a result of these and other factors, people with an educational background and training in the earth sciences are in great
demand. In a recent issue of *Money* (May 1980, vol. 9, p. 76),
geology was listed as one of the four "sunniest" occupations, primarily because of increasing domestic oil exploration. The Bureau
of Labor Statistics has estimated that approximately 2,300 jobs
for geologists will open up each year through 1985 while only

TABLE I

**Estimated Percentage Job Growth in the 1980s
(1980–1990)**

Job	Percentage Growth
Mining engineers	58.3%
Geologists	41.3%
Petroleum engineers	37.7%
Marine scientists and oceanographers	21.1%
[Teachers, college and university]	−14.3%

—Data from the Bureau of Labor Statistics

1,500 geologists and other earth scientists are entering the job market in each of those same years. Because the demand is expected to be greater than supply, job prospects for earth scientists are excellent. The predicted growth rate for positions employing geologists, at least through 1985, is expected to be 41.3%—a higher rate than that predicted for any other science or engineering field! (See Table I for a breakdown of the estimated growth. Jobs in teaching are included for reference.) Looking even further ahead, a recent survey (*Chemical and Engineering News,* 8/13/79, p. 7) stated that "serious shortages of chemical engineers, as well as petroleum engineers, geologists, and geophysicists, will continue well beyond the 1980s." Thus, geology and the earth sciences promise to be growing areas of career opportunities for many years to come.

Over 61,000 people already work as earth scientists in the United States. Half of these are geologists, but all branches of earth science are represented (Fig. 1.1). These people, in turn, are part of a worldwide scientific community that includes about 550,000 geologists. Of these, 16.5% are in North and South America. Africa, which contains 22% of the world's land surface, has only 1.5% of these geologists.

Salaries in geology and other earth sciences in the United States vary with experience, educational background, and the type of work. They also vary by type of employer and the availability of geoscientists, but the salary levels are generally high. A sum-

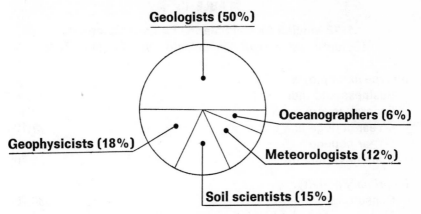

Fig. 1.1   This pie chart represents earth scientists employed, by specialty, as of 1978 (total = 61,000). Data come from *Occupational Outlook Quarterly*, 1980–1981, U.S. Bureau of Labor Statistics. Percentages total more than 100% due to the rounding of figures.

mary of average salaries is shown in Table II. Numbers like these are difficult to compile; they are often several years out of date by the time they have been collected, but they do provide a fair picture of salaries in the geological sciences at the time of this writing. These salaries will continue to increase annually. A survey in 1981 by Vine Associates reported that earth scientists with five years of work experience earned an average salary of $42,800. The range of senior-level specialists, with ten years of experience or more, ranged from $38,100 to $68,600. The employment area in which salaries are subject to the most change is industry, particularly petroleum companies. At the 1981 meeting of the American Association of Petroleum Geologists, for example, salary offers to college graduates with no experience averaged $26,000. The Scientific Manpower Commission has reported that in January 1982 geological engineers with bachelor's degrees were being offered jobs in industry at annual salaries of almost $27,000—an increase of more than 10 percent over the previous 6 months. Geologists with master's degrees received average offers of nearly $28,000 per annum.

A survey by the American Geological Institute, published in

TABLE II

1978 Median Annual Salaries for Earth Scientists
(Scientific Manpower Commission, Washington, DC)

By type of employer:

Business and industry	$33,600
Federal government	28,400
4-Year college and university	27,100
2-Year college	26,400
State government	20,700

By primary activity:

Consulting	$36,700
Management and administration (non-research and development)	36,500
Management and administration (research and development)	35,700
Basic research	30,300
Applied research	30,200
Teaching	26,600
Other	26,800

December 1981 but using 1979–1980 salary information, gave additional examples of starting salaries in industry. Offers to geologists with bachelor's degrees ranged from $11,000 to $33,000, with 44 percent between $18,000 and $26,000. Most (78 percent) of the offers to geologists with master's degrees ranged between $22,000 and $35,000, and nearly all (92 percent) of the offers to PhD's were between $24,000 and $36,000. Salary offers for college-level teaching showed a wide range; at the other end of the spectrum, offers to earth scientists with bachelor's degrees to teach secondary school ranged from $7,500 to $12,000. Such low salary offers may help explain why there is such a shortage of qualified earth-science teachers. According to a 1981 survey by the Education Department of AGI, the state of Texas alone needs nearly 1,300 certified earth-science teachers just for eighth-grade middle-school classes!

Clearly, if you want to become an earth scientist, you should have no trouble finding a job and making a good living. Unlike many fields in this age of cutbacks and layoffs, the earth sciences are growing, As G. M. Friedman stated in an editorial in *Science* (1978, v. 201, p. 215), this is "the golden age of the geoscientist."

Earth Science and Society

As the modern world gets more complex, the earth sciences become more important to society. The problem of finite resources is just one aspect of this change. There are limited quantities of minerals, fossil fuels, and water, but a growing population and increased industrialization mean increased demand. Additionally, the problem of unequal distribution of these resources among countries throughout the world has led to a new understanding of the age-old relationship between geology and political science—geopolitics. The control of valuable resources by a nation gives that country political leverage in world affairs. Everyone is aware of the power of OPEC members (Organization of Petroleum Exporting Countries) but oil is only one of these levers. There are other resources that are found only in politically unstable areas. Manganese is another example of this; the United States imports 98% of the manganese it uses, primarily from Gabon, Brazil, and South Africa. The importance of this mineral was underlined by the U.S. Geological Survey, which stated: "When we can do without steel, we can do without manganese."

Besides the problems associated with finding certain resources, there are also growing concerns over the disposal of industrial waste products. Chemical and nuclear wastes are the inevitable by-products of some manufacturing processes and energy production. Toxic chemical waste now accumulates at the rate of 57,000,000 metric tons per year, and this material must be kept from reentering and polluting the environment. Some wastes have a long lifetime. Nuclear waste, for example, remains dangerous for many thousands of years. Where should these wastes be stored? Any site suggested for long-term storage must be safe and geologically stable over extended periods of time, and geologists are the only scientists with an understanding of earth processes over such long time periods. Geologists are needed to help locate suitable storage locations. Proposals have included burying waste in salt mines or excavated igneous rocks, sending waste on rockets into the Sun, and depositing waste in ocean trenches with the hope that it will be recycled into the earth's interior. There are problems with all of these suggestions, and skilled earth scientists will be employed to help resolve those problems during the next decade.

Land reclamation also requires the expertise of earth scientists who understand processes that operate at the surface of the earth, or "surficial" processes. In the production of minerals and some energy resources (like coal), strip mining is far safer for the miners than subsurface methods. In some cases, surface mining may be the only economic way to extract minerals. However, strip mining disturbs the environment to a far greater extent than below-ground mining. Only a thorough understanding of both the natural processes in the area and human values will allow earth scientists to restore the area to a condition close to its undisturbed state. As mining and large-scale alteration of the earth's surface increases, skilled geoscientists will be needed to plan the reclamation of the damaged land. On a much smaller scale, environmental landscaping also uses the skills of earth scientists who understand surficial geologic processes. Because the natural limitations of the land define the favorable land uses, planning must begin with geology. The availability of water is also important in land-use planning; the precipitation, stream discharge, drainage areas, and slopes all affect how suitable a location will be for construction.

Soil provides support and nutrients for plants, and therefore its conservation is essential to the world's food supply. Soil develops through the weathering of rock materials and the accumulation of organic matter; decaying plants produce an acid (humic acid), which further enhances soil development. Many years are needed for the formation of soil that can grow crops, but the soil can be removed or ruined rapidly. Erosion by wind and water, depletion by agricultural overuse, and excavation for roads, housing developments, shopping centers, or industrial development can all eliminate much-needed farm land. The importance of soil as a natural resource was dramatized by massive losses of the 1930s Dust Bowl days. Also, increasing awareness of world food shortages has renewed interest in soil resources.

Planning proper land use is a major aspect of modern earth science. All land use is fundamentally dependent upon the geology of the area. Geologists and earth scientists seldom actually make decisions concerning land use, and therefore they need to provide environmental information to decision makers clearly and con-

Doing field work outdoors is a fundamental part of all earth science. These two geologists are studying the structure in the Coast Ranges of California. (Photo by the author.)

cisely. By presenting geologic information in a way that is understandable to non-scientists, whether through public lectures, legal testimony, or newspaper articles, geologists can help planning boards and government officials (as well as private citizens) make wise decisions regarding the use of land and natural resources.

Clean water and air are also becoming more valued resources. Although the earth has abundant supplies of both these commodities, their quality is a continuing concern. Contamination of rivers and groundwater supplies and widespread air pollution are problems that need to be dealt with—and solved—by earth scientists.

Air pollution may be the result of natural topographic conditions, as well as excessive industry. The smog in Los Angeles is primarily a result of the ring of mountains that surrounds the city and onshore winds that trap air in the topographic basin created by the mountains. Recently, other cities like Denver have begun to have problems with air pollution. Another atmospheric pollution problem, acid rain, created by corrosive emissions from industries, is being carried across city and state boundaries to damage vegetation in other areas. In fact, Canadian citizens have complained that acid rain from the United States has been having damaging effects on their forests and farmlands.

Relationships between geology and human health are now becoming better understood. The association between a lack of iodine and thyroid disorders in humans has been understood for many years, but other relationships are only now being discovered. Connections between hard water (with many dissolved minerals) and heart disease, the various trace elements that affect human health, and even the influence of local geology on soil and agricultural products all now seem to be more important than previously suspected. Geologists and geochemists, working with biochemists and nutritionists, will help us lead healthier lives in the coming years.

As our technology improves, geologic exploration is expanding beyond our own planet. The 1970s saw a geologist, Harrison Schmitt, collecting rock samples on the surface of the moon. In the 1990s, a sample-return mission may bring back samples of the Martian regolith (the rocky "soil" that was created by meteorite-impact fracturing of the surface materials). Until Mars can be examined directly, however, planetary geologists must rely on a variety of remote sensing techniques to infer the nature and characteristics of our neighboring planets. These studies are valuable both in increasing our understanding of the Solar System and also for the ways in which they illuminate our knowledge of the origin and geologic evolution of the earth and its atmosphere. For example, the remote sensing techniques that have been developed to study other planets are now helping us study our own planet.

One of the ways that space exploration has helped us understand our own planet involves our "sister" planet, Venus. Even though Venus is nearly the same size as Earth and not that much closer to the sun, the atmosphere of Venus is nearly 100 times as dense as Earth's, and the surface temperature is over 450°C (almost 900°F)! Why is it so hot? Atmospheric scientists believe that the answer may lie in the high percentage of carbon dioxide in the Venusian atmosphere. This has important implications for the future of Earth: the burning of fossil fuels increases the carbon dioxide content of our own atmosphere. Carbon dioxide absorbs heat that the earth normally radiates into space. When the amount of carbon dioxide in the atmosphere increases, the amount of trapped heat increases, thus raising the temperature of the earth's atmosphere. Because the earth's climatic system is so delicately balanced, an atmospheric temperature change of only a few degrees could begin an ice age—or melt the ice caps and flood hundreds of coastal towns and cities. (An increase of the earth's average temperature of 1°C would make the global climate warmer than it has been at any time within the last 1,000 years.) Other geologic phenomena may also alter our climate, including ash and dust associated with volcanic eruptions, minor variations in the earth's orbit, and movement of the continents (continental drift, as described by the plate tectonic theory). These events have a significance beyond climatic change alone. With a growing world population, food production becomes a subject of increasing importance. Understanding the processes that affect the climate may help us both to predict and to control climatic changes.

The Global 2000 Report to the President, published in late 1980, presented a rather gloomy outlook on what the earth might be like in the year 2000 (see box). Rather than being a cause for pessimism, however, this outlook should be viewed as a challenge. The Global 2000 Report describes the way the world *could* become, if present trends continue unchanged—not the way it has to be. The report indicates areas of needed research, including a better understanding of the relationship between atmospheric carbon dioxide and global temperature, the environmental effects of impurities released by coal mining and burning, and the reasons why mass extinctions of both plant and animal life occur. Geologists and other earth scientists are the people who understand the fundamental problems facing our planet—and they are also the ones who will help solve these problems in the years to come.

FROM THE GLOBAL 2000 REPORT TO THE PRESIDENT

The total production of new mineral resources each year is now equivalent to 20 tons for each citizen of the U.S.

If this were true for the entire population of the developed world, the total consumption would be about 2×10^{16} grams per year—about the same amount of crust that is formed at the ocean ridges annually.

Energy demands will increase almost 50% between 1980 and 1990.

Most metals and hydrocarbons are being put into the oceans at a rate 2 to 100 times faster than they were before the influence of man.

By 2000, 20% of all species (both plants and animals) living on Earth now are predicted to be extinct —perhaps as many as 2 million species.

CHAPTER 2

What Is Earth Science?

As it has been well said, not one of us, if we are scientists at heart, can afford to ignore any branch of our science, "even though it be conspicuously—and even glaringly—useful."

—Charles Lapworth, *Geological Magazine* (1899)

Science is the orderly study of a subject, and geology and earth science both involve the study of the earth. The earth, in turn, is composed of many parts: the hot interior, the thin crust near the surface that contains minerals and petroleum, the soil, water, the atmosphere . . . the list may not be endless, but it it very long indeed. Earth science is the general term describing a group of sciences that concentrate on these individual aspects. The major branches are geology, geophysics, engineering geology, soil science, oceanography, and meteorology. As shown in the previous chapter (Fig. 1.1), geologists account for half of all earth scientists, but all the branches of earth science are important.

Geology (from the Greek *geo*, meaning earth, and *logos*, meaning study) is usually divided into two parts, physical and historical. Physical geology deals with the materials and processes of the earth, from the hot interior of our planet to the rivers that erode rocks and sand grains at the surface. Historical geology is concerned with geologic materials and processes through time, including the origin and evolution of life. Because our planet is now calculated to be nearly 5 billion years old, historical geology includes considerable time and space! The divisions between physical and historical geology are often loosely defined, and many geologists use processes that can be observed today, as well as evi-

13

dence preserved in the geologic record, to study a particular process or feature.

One example of the variety of approaches geologists use toward a single problem can be seen in sedimentology. Sedimentologists study the erosion, transportation, and deposition of rock particles (sediments), which may range in size from dust to boulders. The agents that act on these particles include wind, waves, rivers, and chemicals. To understand the processes fully, a sedimentologist might travel to Libya to study modern sand dunes in the Sahara Desert's Ubari Sand Sea, and to the American Southwest, to look at outstanding exposures of the Navajo Sandstone, which includes ancient dune deposits that formed about 150 million years ago. Then, for a broader perspective, the sedimentologist might analyze some of NASA's photographs of the huge dune fields in the polar regions of Mars. Thus, trying to understand a single type of geologic feature has taken our sedimentologist to Northern Africa, the southwestern U.S., and another planet!

Earth science began as an academic discipline known as "natural philosophy." In fact, many geology departments in older American colleges and universities began as departments of natural philosophy. Geology and the other earth sciences began to establish themselves as independent sciences in the nineteenth century. However, the shift toward quantification and numerical analyses —and away from simple description—allowed the geosciences to be accepted as full-fledged members of the scientific community. [Soil science is going through a similar transition into quantification now.]

The development and acceptance of the theory of plate tectonics in the 1960s demonstrated the power of the earth sciences to explain natural processes. This theory postulates that the earth's crust is divided into eight large plates (and numerous smaller ones) that fit together like a gigantic jigsaw puzzle and float on more fluid materials deeper in the earth. The plates jostle and bump against each other, causing earthquakes at the boundaries between them. Although a more comprehensive theory may replace plate tectonics some day in the future, this model now helps geoscientists explain such diverse phenomena as earthquakes, the locations of mineral ore bodies, the ages of the ocean basins, and locations of volcanic activity.

The concept of geologic time is the essence of geology. An

The basic scientific tools of observation and description do not change with time. Here, geologist A. H. Brooks is shown examining a mine in the Seward Peninsula region of Alaska in 1900. (Photo courtesy of the U.S. Geological Survey.)

appreciation of the extent of geologic time, over 4.5 billion years, is probably the greatest contribution of geology to Western thought, and it is also what distinguishes earth scientists from other physical scientists. This thought was well stated by geologist Adolph Knopf, who said, "If I were asked as a geologist what was the single greatest contribution of geology to modern civilized thought, the answer would be the realization of the immense length of time. So vast is the span of time recorded in the history of the earth that it is generally distinguished from the more modest kinds of time by being called 'geologic time'."

Most of the branches of earth science overlap with several others; few can be defined by neat distinctions. Examples and brief descriptions of some of the major specialties within geology and earth science are given in this chapter. As is true in any field, each of these subdivisions can be subdivided still further. The discussion here is only a brief introduction to the many aspects of earth science. It is important to remember that all earth scientists have some knowledge about specialties besides their own.

Geology and Geophysics: From Sand to the Stars

Mineralogists study the fundamental material of the earth— naturally occurring chemical compounds known as minerals. Mineralogists are concerned with the origin and composition of minerals, their properties, their occurrence, classification, and use. Groups of minerals form rocks, which are studied by *petrologists*. Petrologists study how rocks form, their mineral composition, and structures within the rocks that offer clues to their formation. This subject area can be subdivided further by the types of rock that are studied. *Igneous petrologists* are concerned with the formation of rocks from magma (molten rock material), both on the earth's surface and deep within the earth. *Volcanologists* specialize in studying eruptive activity on the earth's surface, both ancient and modern. *Metamorphic petrologists* are concerned with the alteration of rocks by combinations of intense heat and pressure, including the new minerals that are produced. *Sedimentary petrologists* investigate the processes by which sediments, small particles of old rocks and minerals, become new rocks. *Sedimen-*

tologists also study sedimentary rocks, but they are more interested in how the sediments were eroded, transported, and deposited than how these sediments were transformed into rocks. *Petroleum geologists* use both field work and geophysical techniques to search for oil and natural gas within the earth. *Geochemists* study the composition of rocks and minerals to determine their origin, distribution and variations. They also calculate the ages of rocks using the chemistry of the radioactive elements they contain.

Stratigraphers investigate geologic history by studying the relationships among layers, or beds, of rocks. For example, using the principle of superposition, stratigraphers know that the rocks on the bottom of a pile (or stratigraphic column) will be the oldest, and the youngest rocks will be on top (unless the whole sequence is overturned). Rock strata can be interpreted in terms of their origin, thickness, composition, fossils, age, and overall history. The strata can then be correlated from place to place around the earth. *Structural geologists* try to explain the orientation of rock units, using faults, folds, cracks, and stretched or compressed fossils to explain the stresses that have affected an area over a period of time. *Paleontologists* study the origin, evolution, and sometimes extinction of organisms—including Precambrian microorganisms, algae, shellfish, flowering plants, dinosaurs, and man. These studies all contribute to the work of historical geologists, who use landforms, structure, stratigraphy, paleontology, paleomagnetism, and various radioactive dating methods to reconstruct the evolution of the earth and its atmosphere from its origin to the present day.

Environmental (or engineering) geologists provide valuable information to planners and decision makers concerning suitable uses for particular areas. These geologists are concerned with the environment, including three basic topics: resources, hazards, and planning. Resources include the location and amount of minerals, energy resources, and water, as well as the environmental impacts of extracting and using these. Environmental hazards include floods, landslides, earthquakes, volcanic activity, and other geologic events that can be dangerous to man. The natural limitations of the earth, in both hazards and resources, are reconciled with assets and social needs by planning.

Landforms on the earth's surface and the processes that form them are studied by *geomorphologists*. The processes range from

Modern laboratory work in earth science involves complex equipment, like this atomic absorption spectrophotometer, which helps scientists determine the concentration of elements in water-quality studies. (Photo courtesy of Jay Fussell, University of Nebraska.)

scouring by glaciers to sand movement along a beach. Water is the most important geomorphic agent, even in arid regions, and those who specialize in flowing water on the surface—rivers— are called *fluvial geomorphologists*.

The practical, or applied, areas in earth science include engineering and economic geology. *Engineering geologists* apply geologic information and techniques to earth materials in order to evaluate their use by man. The uses for this information may include locating a dam, building a road, or finding sources of groundwater for a new city or industry. *Economic geologists* concentrate on the rocks and minerals that can be used profitably by man; these materials can be as exotic as gold and diamonds or as commonplace (but no less important) as sand and gravel.

At the other end of the spectrum, *planetary geologists* (also called astrogeologists) apply most of the specialties described above to other planets and their satellites. So far, man has done "field work" on only one place besides the earth, during the Apollo missions to the moon, but many thousands of spacecraft photographs of other planets have been sent back to Earth. By the 1990s, samples from the surface of Mars may be returned, to be studied in laboratories on the orbiting Space Shuttle and on the earth. Some suggest that, one day, we may find geologists in space, mining asteroids and the moon for their valuable raw materials. Some of this material may be returned to Earth, but much of it will be used for construction in space or on other planetary bodies. Using the resources available in space will be an important aspect of solar system exploration in the years to come.

Solid earth geophysicists apply theoretical and exploration techniques to study the composition and structure of the earth, both near the surface and at great depths. Seismic prospecting techniques use sound waves that bounce off rock layers to evaluate the possibility of accumulations of oil or natural gas. *Seismologists* concentrate on natural and man-made earth vibrations to understand subsurface structures, the movement of land masses, and stresses within the earth. Some geophysicists specialize in the theory and practical application of the earth's gravitational and magnetic fields. *Geodesists* make precise measurements and calculations to determine the shape and size of the earth and local and regional movements of the earth's crust.

Oceanography and Meteorology: Water All Around Us

Hydrologists study water on or in the continents, often concentrating on subsurface or ground water. Hydrology includes the study of water that falls as precipitation and water being discharged from the continents into the ocean. Once water has entered the ocean, it joins the realm of the oceanographer, who studies the chemical, physical, biologic, and geologic aspects of the ocean. Because nearly three quarters of the earth's surface is covered by water, oceanographers have a large area to study. *Chemical oceanographers* evaluate the chemical composition of ocean sediments and sea water, as well as reactions that occur in that environment. *Physical oceanographers* concentrate on the physical characteristics of the ocean, including currents, tides, waves, and physical properties of water masses.

The relationships between the oceans and the atmosphere may result in more accurate prediction of the weather. However, *meteorologists* do far more than simply make weather forecasts. *Physical climatologists* are concerned with the properties of the atmosphere, including snow, rain, clouds, and the chemistry and physics of the gaseous envelope that surrounds our planet. *Synoptic meteorologists* use current weather information to predict short- and long-range weather conditions. *Climatologists* use weather observations collected over long periods, perhaps many years, to study trends in local and global climates. As the world's population grows, the threat of a famine in any part of the world becomes a more frightening prospect, and the understanding of long-term climatic changes may play an important role in preventing a global disaster.

Soil: The Basis for Life

Soil scientists study the physical and chemical nature of soil. Because nearly all plants depend upon soil to grow (only a few plants, like Spanish moss, survive on atmospheric moisture), life on Earth is fundamentally tied to the existence and quality of the soil. Soil scientists may study irrigation techniques or erosional processes, evaluate the need for fertilizer or chemical additives to stimulate plant growth, or consider various agricultural processes. The work of these men and women is often carried on at a local level, working with the individual farmers and land owners.

The problems that soil scientists work with are vitally important to the entire country, even though the average urban resident may never think about it. In the last 100 years, the state of Iowa has lost 50% of its topsoil; for every bushel of corn harvested in a year in Iowa, two bushels of topsoil are stripped away. New farming techniques may help save some of the remaining soil, but the losses will take many years to repair. The conflict over the proper use of soil as a natural resource is exemplified by the problems associated with the production of "gasohol." Plant material can be used to create this substitute for fossil fuels . . . but only at the expense of valuable soil that could also be used to grow food. Because soil is of such vital importance, it is no surprise that most soil scientists are employed by the federal, state, and local governments.

Soil science is undergoing a revolution similar to the one that geology went through in the 1960s as it changed from a qualitative, descriptive science to a quantitative, more precise one. This shift in soil science is a trend toward more consideration of processes rather than dealing in static, unchanging relationships. Soil scientists are becoming concerned with fluxes and flows— with the energy and materials entering and leaving a system, rather than simply the nature of the soil at any given time.

Soil scientists are also in the midst of expanding the time scale on which they think, and making their time perspective much closer to that of geologists. In the past, they have viewed processes in soil science on time scales of a few years, reflecting the agricultural crop cycle. Now, they are becoming more concerned with longer time scales, up to 1,000 years or more; these lengths of time are significant in terms of chemical transport of material through the soil. The long-term effects of modern activities by man, especially those related to soil pollution, are also being studied. As these time scales expand, the relationship between soil science and geology becomes more direct.

What Do Earth Scientists Do?

One of the advantages of a career in geology and the other earth sciences is that it is seldom dull or routine. Any "typical" earth science job may involve a wide variety of tasks, including collecting data in a remote part of the country (or world), analyzing the collected samples with sophisticated laboratory techniques or high-speed computer programs, and writing scholarly reports on the results. Many branches of earth science have aptly described themselves by saying "The world is our laboratory," and there is truth in this statement. Earth scientists can seldom perform experiments on nature in the same way that chemists can study solutions in test tubes or physicists can analyze physical forces with

After collecting data in the field, geologists compile the information into geologic maps. Office, laboratory, and library work complement the field work; geologists work in many different settings. (Photo by the author.)

equipment. The earth sciences demand a special creativity. For example, when stratigraphers and sedimentologists study the ancient geologic past, they cannot travel back in time and do experiments to see what the composition of the atmosphere was several billion years ago, or what the temperature was, or where the continental land masses existed relative to each other. The information must be collected in less direct ways, using every available method. Paleontologists can use the fossil record (if any) to infer the organisms that lived in the past and something about their environment; sedimentology may reveal something about the processes that were active on the earth's surface; geochemistry can tell us about the available trace elements; and so on. Often, ingenious methods turn out to be the most useful—and the most controversial.

Thus, earth scientists perform many tasks in the process of doing their jobs. Some of the major activities will be discussed here, but the best way to get an idea of what a job involves is to talk to someone who does it for a living. You'll find most earth scientists are delighted to talk about their work, and some may invite you to visit them in the field, laboratory, or office for a first-hand look at what earth scientists do.

FIELD WORK

Almost all geologists, soil scientists, oceanographers, and other earth scientists spend some time doing field work, even if it is only for an occasional summer. Most continue doing some field research throughout their careers.

Many studies in the earth sciences, and particularly in geology, involve research in a specific geographic area. This may be as small as a single outcrop of rock or as large as an entire state or country. Geoscientists speak of being "in the field" or "going into the field" when they are going to be working outdoors in their study areas. Life-styles while doing field work vary with each individual and the available facilities—field work is often done in remote, primitive areas. Camping and cooking out may be necessary, or small motels and local diners may become a "home away from home." Some field areas can be reached only by backpacking (a difficult job if heavy equipment is needed for the study, or if many rock samples must be collected) or by four-wheel drive vehicles. Other areas, especially in mountainous regions such as Alaska or the tropical forests of South America, are virtually impossible to study without helicopters.

If you don't like the possibility of sleeping on hard ground or strenuous climbing, you might consider one of several other earth science fields, such as chemical oceanography or meteorology, as an alternative career. These disciplines, however, also require some field work of a different kind. Oceanographers must sometimes collect samples at sea, on research ships that may spend months between ports. Meteorologists may collect data on a daily basis; with increasing computerization and remote sensing, fewer direct readings must be taken, but no substitute has been found for individual observation.

While doing field work, many scientists work in teams; this is a useful arrangement for many reasons. There is no substitute for actually seeing a field area in person, and having another person share his or her thoughts with you while you are studying a particular area is more valuable than discussing it "back in the office." Having a partner or field assistant can also be an advantage when taking measurements, especially across any distance. On a more personal level, a field partner can provide conversation and an extra hand with the camping, cooking, and carrying.

The safety aspect of teamwork is also important. Although geologists apparently tend to be healthier than average as a group (generally attributed to continuing field work, often many years past retirement), they are naturally susceptible to field-related accidents. Each year, several scientists and graduate students are injured doing their work, by helicopter or airplane crashes, falls from cliffs or ledges, and even grizzly bear attacks. Having a partner can make a big difference in administering first aid and finding help. However, the occupational hazards in earth science are certainly less than those of teaching in most high schools today, and many of these field accidents can be prevented. As a result of a growing awareness of the problem of field-related accidents, some graduate schools are beginning to offer first aid courses to their students.

Working conditions in the field can also vary with the type of work being done. Mining engineers and mineral-production personnel may find themselves working hundreds of feet beneath the earth's surface. Meteorologists and atmospheric scientists may be required to take measurements from high-altitude aircraft. Working on offshore drilling rigs can involve particularly hazardous conditions; the fatal-accident rate on rigs drilling in U.S. waters is three times higher than the accident rate for coal miners or building construction workers. Being run into by a ship is always a concern on offshore platforms, and the danger of a blowout, which can occur when the drilling bit penetrates a zone in which fluid is under high pressure, is a constant threat. Besides being a threat to worker safety, blowouts can also result in environmentally damaging oil spills, like the one in the Santa Barbara channel in 1969. Offshore drilling rigs operate 24 hours a day, so crews may find themselves working 8- or 12-hour shifts in the middle of the night.

Scientists sometimes find themselves working in extremely isolated circumstances, like offshore platforms, oceanic research vessels, or remote observation stations (in the Antarctic, for example). In these situations, considerable emphasis is placed on keeping the workers happy; the food is usually excellent, and a variety of movies, tapes, and games are available during off-duty hours. Employees who take this type of duty may be compensated by larger paychecks, more vacation time, or later transfers to more desirable locations.

The most remote field area studied so far is the moon. The collection, return to Earth, and subsequent study of lunar rock samples led to a dramatic increase in our understanding of the solar system. Within the next 50 years, "field work" on other planets, particularly Mars, is a reasonable possibility. Samples of martian dust and rocks may be collected by astronauts or sophisticated robots; either way, analyses of the material will lead to major advances in planetary science.

LABORATORY RESEARCH

Top scientists generally have assistants to perform routine lab work for them, but almost everyone starts out doing his or her own laboratory analyses. Following the necessary procedures is often time-consuming, but it is an invaluable lesson in understanding the methods that lie behind analytical results. Some people prefer to do their own lab work throughout their careers because it is the only way to control the procedures and have confidence in the quality of the results.

Many laboratory analyses involve geochemistry: determining the composition of a particular rock or mineral, studying trace elements in water or soil samples, or studying the remains of ancient plants or animals. Clearly, a solid background in chemistry is useful. In fact, when it comes to geochemical lab work, understanding chemistry may sometimes be more valuable than knowing geology; the interpretation, however, necessarily relies heavily on geological understanding.

Microscopes are important pieces of laboratory equipment. They range from the simple hand lens that the field geologist carries in his pocket to the Scanning Electron Microscope (SEM) that micropaleontologists use to study tiny organisms that may have been important in the formation of oil. Between these two extremes is the petrographic microscope, which is used to study rocks in "thin section." These are extremely thin slices of a rock sample, only about 0.001 inch thick; individual minerals display unique characteristics when polarized light is transmitted through them, and the mineralogical composition and texture of a rock can be studied in great detail using thin sections.

PRESENTATION OF RESULTS

The day of the reclusive earth scientist, interested only in his or her own studies and field work, is over. Scientists in many fields have come to recognize the importance of being able to present scientific results, both orally and in writing. No one knows how many brilliant insights have been lost because their authors failed to express them clearly and concisely!

Speaking skills are important communication tools in the earth sciences. One of the factors contributing to this is the slowness with which full-length scientific papers are published now. An average waiting period for getting a paper published in a scientific journal is a year—or more. If everyone waited until he or she could read papers that reported important results, the progress of science would be very slow indeed. Because of this,

We do much of our thinking with words, and if I have trouble understanding what a geologist writes, I wonder whether he or she *thinks* clearly and logically. That is, if he writes badly, does he think badly? Of course it may not matter; I know geologists who are good scientists but poor writers—but the bad cancels the good. Clumsy wording can mean you won't get the job, the company won't drill the well, the dam will be built in the wrong place, or your boss will delay your promotion.

Wendell Cochran
Editor, *Geotimes* and *Earth Science*
Coauthor, *Geowriting* and *Into Print*

scientific meetings are becoming more important. Recent results can be presented orally or by written abstracts (short summaries) in formal sessions—and informally during the coffee breaks and evenings. These reports allow scientists to find out the most recent discoveries in their fields. Individual specialties are often so small that a scientist can have at least a nodding acquaintance with most of the other people working on similar problems. Often, the best opportunity to exchange ideas is at a professional meeting.

A number of schools are now beginning to require students to make oral presentations of the results of projects that they have worked on. This may even take the form of a simulated professional meeting, which may include a system of blinking red lights that indicate the allotted speaking time is over or a formal question-and-answer session with the audience. Although these sessions may seem mildly traumatic for the students at the time, the experience is important. Such presentations will help the speakers feel more comfortable when they address other groups, and it is generally a mistake to shy away from opportunities to practice one's communications skills.

Besides reporting results to their colleagues, scientists are often called upon to explain their work to nonscientists. For example, a geologist might give a presentation to the officers in his or her company, trying to convince them to explore a new area, buy a certain piece of land, or bid for drilling rights. Earth scientists may find themselves working with city planners, testifying in courtrooms, or participating in committee hearings on environmental problems. In these situations, writing and speaking abilities are extremely important, in order to make complex technical problems understandable to people unfamiliar with science.

Writing ability may be even more important than speaking ability; publishing a paper in a scientific journal means that, generations hence, other scientists can still read your work. You wouldn't want them to think you couldn't write or spell! The difference between a poor idea and a great idea that is poorly expressed is often hard to distinguish. Well written papers are generally more likely to be accepted for publication, and if the reader can understand it, he or she is more likely to be convinced by your arguments. If a paper is well written, the worst a critical colleague can say is "I don't agree with what you say, but at least I understand it!"

Writing clearly and concisely is also important for economic reasons. Every word costs money to edit, typeset, and print. Publishing costs have risen dramatically in recent years, and some journals are now giving preference to shorter papers by publishing them more quickly than extensive monographs.

Communications skills include the ability to deal with people. Besides the communication of research results, this means cooperation with colleagues, especially in the field. An abrasive field partner usually cannot be avoided, especially in an isolated area, but you can simplify your life if you have learned to handle a variety of personality types.

All earth science is a form of teaching, whether you are working with students, supervisors, or clients. Transmitting knowledge is part of the job of being an earth scientist, and your work will be a lot easier if you are well trained in the skills needed for success in this area.

TRAVEL: THE MEANS TO AN END

Earth science cannot be done exclusively in labs; earth scientists must sometimes go to where the rocks, soil, water, or weather are. They collect samples and study them further in the laboratory, but neither the field work nor the lab work can exist without the other. Thus, travel is essential to earth science. In fact, job opportunities may be limited if an applicant is unwilling to travel at all.

The required travel may be as limited as a few overnight jaunts per year or as extensive as a year—or more—working overseas. Oceanographers often have shipboard scientific duties that take them away from home for three or four months or more at a time. An invaluable asset here is a supportive family, and the willingness to move may be an important attribute in a husband or wife.

The problems of long absences and moving, along with the seemingly endless nights of motels and diners, are offset by meeting new people and seeing new places, along with all the new experiences. Many geologists pride themselves on feeling equally at home in, for example, a small oil town in Oklahoma, a gold mining camp in California, snorkeling off the Florida Keys, or in the executive meeting rooms of an oil company in Houston.

THE GREAT OUTDOORS

It is probably safe to assume that most accountants do not spend their vacations doing math problems, and dentists do not voluntarily look into peoples' mouths when they are off duty, but an astonishing number of earth scientists have "hobbies" that sound very much like what they do for work. Camping, hiking, fishing, sailing, and traveling are often high on the list of an earth scientist's leisure-time activities. This is probably a good indication that they are happy with their choice of career. Many people are drawn to the earth sciences because they love the outdoors. In fact, some colleges have memberships in their geology clubs that far exceed the number of students taking geology courses— because the geology club sponsors the hiking, mountain climbing, and spelunking (cave exploring) outings!

A few branches of earth science, however, involve only limited work outdoors; theoretical geophysics and meteorology may be studied exclusively with computers. People who work in these areas, however, are often no less interested in spending time outdoors than the field geologists!

Geologists are scientists who are experts in their particular specialty as well as generalists who draw on other sciences for their information and insights. The stereotype of a geologist as a rock-hound is no longer valid, if it ever was true. Modern geology has new sophistication and new significance as a profession and as a career. The current search for resources and innovative ways to extract them offer opportunities for discovery. Increasing man's understanding of the earth can be pursued in a laboratory filled with high-technology equipment, on a mountain top of bare granite, along a beach with crashing waves, and in a Space Shuttle 600 miles above the earth's surface.

In the past ten years, geology has gone through a revolution, changing theories to accommodate new information as it has become available. Using a wide range of techniques and a variety of data, geoscientists have evolved ideas about the physical world that are independent from other sciences; one example of this is plate tectonics, which seems to combine a large array of information into a comprehensive theory. Meteorology and soil science are undergoing similar dramatic changes today. This truly is the "golden age of the earth sciences."

CHAPTER 3

Careers in Industry: Energy

Resources are like air, of no great importance until you're not getting any.

—Anonymous

In 1962, geophysicist M. King Hubbert predicted a growing dilemma in our use of oil. Production, he stated, would not be able to keep up with the increasing demand, and we were heading for a major energy crisis. In those days of abundant energy and relative political calm, few people took his warnings very seriously. The international turmoil of the past few years, however, has underlined the finite nature of our nonrenewable energy resources, as well as our need to conserve existing resources and look for new, alternative forms.

There are still Americans who believe that the "gasoline crunch" was manufactured by the big oil companies, but the diminishing reserves of petroleum are very real. We all will have to deal with this problem in the years to come. Nearly 95% of the electrical power generated in the United States is produced from coal, oil, and natural gas. National energy use has doubled every ten years; at this rate, most of the earth's fossil fuels will be completely exhausted within the next 400 to 500 years. Earth scientists are in a unique position of being able to help extract the energy resources that are available and to help find new sources and types of energy, as well as to develop the new technologies to utilize them. All energy and mineral resources are becoming scarcer, harder to find, and more difficult to extract. For example, there are huge deposits of oil shale in the western

United States, but their extraction requires vast amounts of water. In the arid western states, water is a rare commodity, which means oil shale mining is controlled by the technology needed to extract it. The oil exists in the shale; the question is whether we can remove it at competitive cost without damaging the environment and wasting huge amounts of water. Non-fossil energy sources (hydroelectric, geothermal, and nuclear) are being used, but nationally they are not major contributors. The technology for geothermal, wind, solar, tidal, and nuclear fusion power is improving, but definite plans for the future must be made within the context of existing technology.

Petroleum geology and engineering are currently undergoing many new developments. Offshore development is one area that has been advancing rapidly. New areas of interest include the geology of ancient basin areas and faulted regions, where much of the current exploration for domestic oil is going on; new seismic techniques are being developed to help decipher the stratigraphy and structure of potential oil-bearing areas. Additional knowledge of stratigraphy and sedimentation, especially in offshore areas, is needed. These last areas will be particularly important as the next round of offshore drilling leases becomes available for bidding during the 1980s. (Only 670,000 acres, or 0.2%, of the geologically promising areas of the U.S. continental shelf had been leased by the end of 1980.) The opening of federal tracts for leasing in submarine basins off the California coast in the 1980s may provide new opportunities for earth scientists, although the environmental impacts of Pacific offshore production are still controversial. Scientists with the ability to understand the geology of these potential oil-producing areas will be in great demand. Geoscientists working in the petroleum industry will be in the forefront of geology—always an exciting place to be. In the 1980s, exploration of potential petroleum-bearing basins off the coast of China will be one such exciting place.

Career Preparation

The earth science specialties in greatest demand by the petroleum industry are stratigraphy/sedimentation, structural geology, paleontology, and geophysics. People with expertise in other

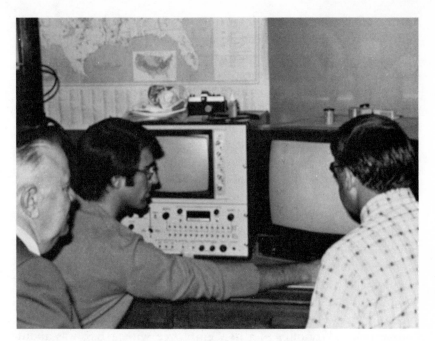

Computer enhancement of photographs can help remote-sensing specialists study the geology, soil types, and vegetation of large areas. With new satellite radar images, the search for oil and gas reserves can be stepped up. (Photo courtesy of Jay Fussell, University of Nebraska.)

fields are also in demand, depending upon the background and qualifications of the individual. For instance, oceanographers, biologists, and engineering geologists can be very useful in planning and obtaining government approval for specific offshore projects. Oil companies do not necessarily give hiring preference to students who have specialized in "petroleum geology"—it is more important that they have a strong background in the areas mentioned above. Most oil companies are looking for educated geologists rather than petroleum technologists.

Experience with computers can be a valuable asset when looking for a job in the oil and gas business, and the same is true for field experience. Working on the production end of the petroleum industry, "sitting" on wells and logging cores (recording information about the core material as it comes out of the drill hole), can also be a foot in the door for other jobs. Some companies offer summer employment to college students, and this summer experience can lead to a job after you graduate. Some companies subsidize field and laboratory research in connection with graduate degrees in the earth sciences; such support can also lead to employment opportunities after graduation.

The optimum educational level for a career in the petroleum industry is a master's degree, but this is not a rigid requirement. Graduates with bachelor's degrees often receive attractive beginning offers; an additional degree is usually needed for advancement, however. The positions in research and development departments of large companies are usually reserved for geoscientists holding PhDs. With time, as more people entering the job market hold advanced degrees, the demand for employees with education beyond the undergraduate level will increase. However, at times when the petroleum industry needs many geologists and geophysicists, an applicant with a bachelor's degree may be able to command a high salary. As noted in Chapter 1, at the 1981 meeting of AAPG (American Association of Petroleum Geologists) the average salary offer to a new employee with a B.S. was about $26,000.

The petroleum industry is an actively expanding field for employment opportunities. Working for industry has many advantages—opportunities for travel, a chance to keep abreast of current techniques, and excellent promotions and salaries.

Two basic responsibilities of geologists within the petroleum industry are finding hydrocarbons and developing or exploiting known resources. The first, finding oil and gas, is the responsibility of the "exploration" geologist; the second, exploiting known reservoirs, is the responsibility of the "production" geologist. Both types of geologist work with many of the basic tools of the trade to perform their tasks—including the primary duty of making geological maps. Since the maps deal with subsurface strata, there is a need to rely on such tools as geophysics (seismic interpretations), combined with data from previously drilled wells for control.

In exploration, prospects are evaluated on the basis

of the maps and potential reserves of hydrocarbon. "Wildcat" wells are chosen to be drilled in unexplored areas to test potential reservoirs. Once the prospect is determined to be economical, the field is turned over to production geologists for development.

In addition to using geological tools, a petroleum geologist has to work closely with other petroleum-related disciplines. Wells have to be drilled, platforms need to be set, economics of the field must be calculated, and basic operations have to be planned. So there is a need to communicate with your counterparts in other disciplines, especially engineering fields.

Since petroleum can occur in varied rock types and various locations, there are opportunities to travel and live in exotic (and sometimes not so exotic) places around the world. I have enjoyed traveling to Indonesia and plan an extended assignment overseas. I have friends who have been to Norway, England, Columbia, Cameroon, Egypt, Saudi Arabia, and Nigeria, just to name a few wellsite locations and offices. If travel appeals to you, then a petroleum-geology career can offer a chance to fulfill your wanderlust.

Locating areas for future exploration is becoming a difficult task. Therefore, the geologist's job is becoming more critical and is a constant challenge. A career in petroleum geology can be exciting and rewarding.

Judy A. Russell
Production Geologist

Employment Outlook: Moving Up and Over

As a result of the increasing need for more energy, new energy sources, and increased independence from foreign oil, petroleum companies are hiring geoscientists in large numbers right now.

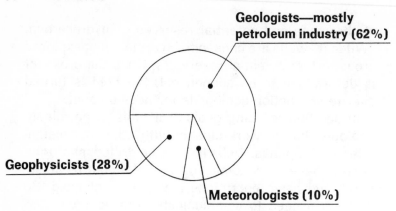

Fig. 3.1   Earth scientists employed by private industry, including petroleum, consulting, engineering, and mineral companies as of 1978 (total = 30,100). Percentages of oceanographers and soil scientists are negligible. Data from *Occupational Outlook Handbook,* 1980–1981, U.S. Bureau of Labor Statistics.

Over half of all earth scientists working in industry are geologists and geophysicists employed by petroleum companies (Fig. 3.1). Prospects in the other energy fields should also continue to improve for a number of years. At the 1980 national meeting of the Geological Society of America (Atlanta, Georgia), Jack Edwards, a former training manager for Shell Oil, pointed out that the petroleum industry had the capability, need, and money to hire 50% of the 6,000 people eligible to enter the job market in geology and geophysics that year!

Common practice in oil company employment is for the "majors"—the major oil companies—to hire graduates by recruiting on college campuses. Over half of their new employees are generally hired as a result of on-campus interviews. These new employees are then given on-the-job training (OJT), along with valuable experience.

Another place where many interviews are conducted is at national meetings of professional societies. Nearly all of the major oil companies interview candidates for petroleum industry jobs at the national (and some local) meetings of the American Association of Petroleum Geologists, the Society of Economic Paleontologists and Mineralogists, and the Geological Society of America.

These interviews may lead to an invitation to visit the corporation office. [The addresses for these and other groups are listed in the appendix.]

Although the pressures of increasing exploration and production of petroleum have increased the number of available jobs in the energy industry, the job market is also inflated by considerable· rivalry within the business. After a few years of experience with a "major," geologists and geophysicists are often lured away by a smaller independent company. In the past, this has been done simply with a higher salary offer. Today, there are increasing numbers of offers that include "production revenue participation" —better known as "a piece of the action." (Independent oil companies can offer these packages more easily than the majors, because they are less restricted by a large board of directors and many stockholders.) One example of the advantages of this arrangement comes from the McMoRan Oil and Gas Company of New Orleans. In 1977, the company lured five scientists away from bigger firms, offering them stock options of up to 30,000 shares apiece. Four years later, all five were millionaires.

This "head hunting" is a widespread practice. By one estimate, half of all the people who are employed in the petroleum business leave within five years for some other job. This is not a new phenomenon—these are just higher rates than ever before. In 1980, the exploration departments of major oil companies lost about 15–30% of their staffs; Shell Oil alone is rumored to have lost about 100 geologists and geophysicists in 1980. Clearly, these vacancies must be filled.

Cycles of growth and retreat are traditional within the oil industry. One of the rarest commodities in the business today is a geologist or geophysicist with 5–15 years of experience. These people fall in the group that became employable during the "bust" part of the last cycle, when oil companies were simply not hiring in the 1960s and early 1970s; imported oil was cheaper than domestic petroleum then, and there was no economic encouragement for exploration and development of oil and natural gas in the United States.

As domestic production became more economical, petroleum companies increased the number of employees, and they continue to do so. The current "boom" part of the cycle also cannot be expected to last forever. The need for traditional petroleum geol-

Geologists and geophysicists on this offshore rig are drilling for petroleum in the Mississippi Canyon, Gulf of Mexico. (Photo by Bill Clark, Atlantic Richfield Company)

ogists and geophysicists will decrease as the nation turns to other forms of energy and as the discoverable resources become harder and more costly to find. These scientists will come largely—if not entirely—from within the ranks of current employees. Consequently, scientists who want to work for oil companies need to be sufficiently flexible to move into other divisions within the companies. Major oil and gas companies have been working toward becoming "energy companies" and "resource companies." As they expand to include most of the resource fields, employees will find themselves able to switch into almost any earth science field. The key to this flexibility is obtaining a broad background that will provide a basis for adapting to different types of energy-related work.

Starting salaries for a new graduate with a bachelor's or master's degree and no experience range from about $21,000 to $26,000 or more per year. After a few years of training and experience, a "lure-away" offer from another company might be in excess of $35,000, plus stock options and other benefits. Because of shortages within the field, 1980 graduates received base salary offers that were 10–15% higher than the offers for 1979 graduates in engineering and geoscience. These increases are dependent in part upon location—in some parts of the country, yearly increases may exceed 20 percent. Estimates of salary levels for experienced petroleum geologists are shown in Table III.

In a 1980 article in *Money* (May 1980) listing the "sunniest" occupations for the next decade, opportunities for geologists were ranked as the fourth sunniest of all jobs. This conclusion was based on four characteristics of geologic employment: estimated job growth to 1990, prospects for job seekers, average starting salaries, and average midcareer salaries. This encouraging prospect was attributed primarily to increased domestic oil production.

The demand for exploration geophysicists is growing rapidly; it is already beginning to strain the supply of qualified geoscientists. One measure of the increased activity in geophysical exploration, the number of people working on seismic crews, increased 32 percent in October 1980 over the previous year. New developments, especially new data gathering and processing techniques, including three-dimensional seismic surveys, have contributed to the effectiveness of exploration. There is a growing realization of how much less expensive a seismic survey is than

TABLE III

1981/1982 Salary Levels for Petroleum Geologists in the U.S.

Years Experience in Industry	Major Oil Companies	Independent Oil Companies
3	$25–27,000	$27–30,000
5	$35–40,000	$45–55,000
7	$50–55,000	$60–80,000*
10+	$65–70,000	$70–100,000+

---

* Participation in addition to base salary (overrides, royalties, etc.) often possible after 7 years of experience.

—Add 5% for graduate degree.

—Add 8–10% for outstanding record of success in finding oil.

Source: Client Search Assignments (in Gaea, 1982, v. 5(1), p. 5).

drilling a "dry hole" (nonproducing well). Information obtained from seismic surveys is now being used for purposes besides oil and gas exploration, including studies to find appropriate sites for petroleum storage and waste disposal facilities, power plants, buildings, dams, offshore oil platforms, and highways. Seismic reflection and refraction surveys and electrical studies are also being used to evaluate hazards associated with potential landslides and subsidence. Stratigraphers who can interpret seismic data are also needed to help explain the results of seismic surveys. A geological understanding of the significance of these raw data is necessary to find practical uses for the information that has been collected.

Related employment opportunities are in the production of oil and natural gas. Most oil companies contract with independent drilling companies to perform the actual exploration and development drilling of oil fields. The foreman on an offshore rig earns around $3,800 per month, drillers earn about $14 per hour, and roughnecks (crew members) make about $7 per hour, with extra for overtime. When a service company is assigned drilling work, the contracting petroleum research company often places a crew of their own scientists aboard to oversee the exploration and record the drilling results.

Working for petroleum companies is not all roses, despite the relatively high salaries. Because of the diminishing resources, geologists may work for many years, drilling only dry holes, with only the satisfaction of having correctly predicted the structure at a location. According to the American Petroleum Institute, less than a third of the exploratory wells drilled in 1980 produced oil or gas. In oil fields, new wells discover oil about 80 percent of the time. Just because oil or gas is found, however, does not mean the well will be profitable. One stock brokerage firm, E. F. Hutton, has estimated that only one out of every 40 or 50 wells is a commercial success. In the last few years, several areas that looked like promising oil fields have turned up dry, including parts of the Gulf of Mexico, Gulf of Alaska, and the mid-Atlantic outer continental shelf. Ultimately, however, the only way to find petroleum is to drill for it. Geologists who have records of drilling successful exploration wells are in great demand and will continue to be.

As a scientist advances in a career in industry, he or she may take on an increasing number of administrative duties. If an individual is good at making decisions, has intuitive leadership ability, organizes projects well, and displays sound judgment, a position with management responsibilities may be hard to escape. In general, management and administrative positions are assigned to employees who have already demonstrated their ability—not to young geologists, with or without a business degree. Later in a career, a business degree may be useful, but geoscientists first need to prove themselves as scientists. If formal training in business will be useful to your company, your employer may be willing to send you back to school.

Many of the places where oil companies are located are less than ideal unless you are particularly fond of heat and humidity. Some of the major sites are Houston, Galveston, and Midland-Odessa (Texas), New Orleans (Louisiana), and Tulsa and Bartlesville (Oklahoma). Somewhat more temperate climates with major oil company offices, however, include Los Angeles, San Francisco, Denver, and Pittsburgh.

Many oil companies have restrictions on publishing scientific results; with the fierce competition for new discoveries, companies are jealous of their data and may be very reluctant to let any information escape to another company. As geoscientists move up the seniority ladder, they are often allowed to do more basic research, and top scientists who do research are often permitted to present their findings—as long as they don't give clues to locations of potential new oil fields! However, most oil producing states and countries have a well-defined procedure through which geological information collected by petroleum companies is made public after a few years.

The Energy Outlook

As fossil fuel reserves decline, emphasis on alternative energy sources keeps growing. With the "incident" at the Three Mile Island nuclear reactor in 1978, concern over the safety of nuclear fission reactors has skyrocketed. A related concern is how and where to store the deadly by-products of the fission process. Nuclear fission involves the splitting of a large atomic nucleus into two fragments, releasing huge amounts of energy and large amounts of nuclear waste. As a potential long-term source, many scientists are turning to research on nuclear fusion. In fusion, the nuclei of two atoms are joined to form a larger one; the nuclear fuel can be extracted from sea water, and the fusion process produces limited radioactive waste. However, the technology for fusion is still being developed. In 1980, Congress passed the Magnetic Fusion Energy Engineering Act, which is designed to have a test facility on line by 1990 and a commercial demonstration plant operating by 2000. There is no guarantee that these will be operating on schedule, however. Clearly, other energy sources will be needed in the interim.

Over the last few years there has been considerable talk about a "bridging" energy source—an energy form that will carry us over into the new technologies of synthetic fuels and nuclear fusion reactors. The most probable candidate for a bridging energy source is coal. Coal has its own problems, however. Burning coal for power adds carbon dioxide to the atmosphere and may cause unknown climatic changes. Underground mining is dangerous and surface (strip) mining damages the environment. However, coal

does have a big advantage—the United States has the largest reserves in the free world!

Coal production has already been stepped up. If the rate of
production increases as much in the next few years as it did in
1980, coal production—and associated employment—will double
by 1990. The need for qualified coal geoscientists can be expected
to grow at the same pace as, or faster than, the need for coal-
generated energy, and this need will be reflected in increased
employment opportunities in both industry and academia. A
number of colleges and universities offer degrees in mining engineering with a special emphasis on coal. A few schools offer two-
year associate degrees in mining engineering, and these can lead
to supervisory jobs. However, most upper-level jobs in the coal
industry, as with other earth science fields, are held by people
with a minimum of a four-year bachelor's degree. Many jobs in
coal production do not require any earth science or even much
educational background, and the pay is good. In 1981, the average
yearly salary for a coal miner in Wyoming was over $25,000. (Coal
mining is no longer exclusively a man's field. Of the 8,500 women
working in the bituminous coal industry in early 1979, over 1,000
were employed as miners!)

Both increased coal production and production of oil from
shale will produce new roles for hydrogeologists. Companies working with coal are planning networks of pipelines through which
crushed coal, mixed with water, will be transported as a slurry.
Oil shale production processes require huge volumes of water. In
both cases, hydrogeologists will be expected to find the water to
supply these needs.

Geothermal energy is an alternative to traditional energy
sources that requires geological, geophysical, and geochemical expertise to locate. Most of the work in this field is being done by
the U.S. Geological Survey, some private entrepreneurial companies, and universities. The methods of exploration and techniques of development have been improving rapidly, and this
results in increased success in geothermal electrical power production. Consequently, economic risks have decreased and should
encourage greater investments (and many more jobs) in the future.

Solar energy is limited only by our ability to convert it into
a useable form. Thus, the technology of solar power will be a
source of employment in the next decade. The environmental

impacts of large solar-cell arrays will require the expertise of earth scientists from many fields. Most of the opportunities related to solar energy will be in small private companies.

Wind and tidal energy are other nontraditional resources under investigation. Wind energy can be expected to meet only a small percentage of the world's energy needs, but it may be a significant source on the local level. As a result, much of the demand for and use of wind energy will be on a small scale. This field will need meteorologists and atmospheric scientists, as well as engineers who understand the technology of deriving energy from the wind. Wind-energy projects are now under way in many places, including northern and southern California. Pilot facilities to harness tidal power have already been constructed in Europe; their development in the United States will require the expertise of both oceanographers and engineers.

Some Points to Consider

A career in energy industries, which are dominated by oil and gas companies, may be right for you if

- You see the importance of energy resources to the growth and development of society.
- You like the idea of sharing company profits.
- You don't object to the frustrations of a low success rate in finding new petroleum sources.
- You wouldn't mind living in a Gulf Coast climate, if necessary.
- You are flexible enough to make the transition to work on alternative energy sources.
- You are ambitious enough to succeed in the corporate structure.
- You are willing to travel, both in this country and to foreign countries.
- You want to be at the forefront of new research methods and technology.

CHAPTER 4

Careers in Industry: Consulting and Engineering

"It is stupid to sleep in the flood plain."

—Don Barnett,
former mayor of Rapid City, South Dakota,
after the 1972 flood.

Consulting companies range in size from one person to multinational corporations. Geological engineering firms are also consulting companies, but they are considered separately because they are so numerous.

Consulting companies work in all the earth science disciplines, and the consultants range from individuals with offices in their homes to members of huge, multinational corporations. The services these consultants offer also cover a variety of tasks, including testing soil for septic tanks, finding suitable locations for nuclear power plants and waste disposal sites, and producing reliable weather predictions for long-distance trucking companies. Many consulting companies provide services to the petroleum industry, and most environmental and engineering companies are concerned with the impact of construction on the environment.

Consultants may find themselves involved in doing research for and writing Environmental Impact Statements (EIS), which are required for most projects involving federal funds. For a large, involved project, the amount of effort and time spent on the EIS can be enormous. For example, the EIS for the Alaska Pipeline filled six volumes and weighed 24.2 pounds!

Environmental and engineering consulting jobs usually in-

45

volve some travel. Projects almost always require a visit to the area being studied and your expenses will be covered by the client. These trips to do field work may be only across town or you may go to the other side of the world. International travel is frequently an opportunity in larger consulting companies. This can be an exciting experience, seeing new countries and cultures.

International travel may sometimes turn out to be too exciting—you may drop in on governments being overthrown or civil uprisings. In recent years, several geologists working in the Middle East have suddenly found themselves in the midst of a civil war!

Overseas travel is one of the few areas in which women may find themselves at a real disadvantage in earth science. Some overseas jobs are done in countries where the culture historically views women differently from men; for example, in the Middle East, where many of the world's major oil fields are found, the traditional cultural view toward women forbids their appearing in public unescorted or unveiled. Naturally, female geologists in their field gear would be strikingly out of place, and both the geologists and their companies must be careful to avoid offending their hosts (and clients), no matter how much they may disagree personally with the cultural view of how women should be treated.

After the research on a project is complete, the information must be presented to the client. Thus, writing and speaking skills are very important in consulting companies. Because the work is done for a client, the results must be transmitted clearly and concisely, free of the jargon that would be permissible in a publication intended for other scientists to read.

A sense of ethics is also important in the consulting business. Because the earth scientists in this field provide the information that major decisions are based on, the scientists' honesty and integrity are critical. Consider the responsibility of siting a nuclear power plant or a large dam: if it fails and many people are killed or property lost (whether through incorrect geologic mapping, sudden fault activity, or some other cause), who is responsible? Even signing a form stating that a piece of land is suitable for development demands a moral judgment on the part of the geologist who signs it.

The constitution of the American Institute of Professional

Geologists installing a creepmeter in central California to measure small movements in the earth's crust.
(Photo courtesy of the U.S. Geological Survey.)

Geologists includes a code of ethics, which begins, "The geological
sciences are a profession, and the privilege of professional practice
requires professional morality and professional responsibility, as
well as scientific knowledge on the part of the practitioner." The
code goes on to describe the importance of integrity in a geologist's relationship with the general public, employers, clients, and
fellow geologists. As might be expected, the penalty for someone
who violates the code of ethics or "conducts himself in a manner
unbecoming a Certified Professional Geological Scientist" is ex-
pulsion from the society.

Some states require the licensing of geologists and engineers

who plan to work as consultants. The trend toward this requirement is continuing as consumer activists exert pressure on state
registration boards to certify professional engineers and geologists.
To meet this trend, several professional organizations (including
the American Society of Civil Engineers and the Association of
Engineering Geologists) offer short courses for geoscientists who
want to obtain or renew their professional registration. In California, the minimum requirement for certification and registration as an engineering geologist is five years of professional
postgraduate experience in the geological sciences. Three of these
must be in engineering geology, but two of the five years may be
graduate studies.

Some states, including California, require "site certification"
before construction is permitted in a given area. As the world's
population grows, the demand for living space creates new pressures on marginal building sites. Flood plains, fault zones, and
areas susceptible to landsliding seem to be attractive, especially if
these areas have features that make them desirable for other
reasons. There are literally hundreds of houses in southern California that are built on unstable slopes and near active fault zones
. . . but with a commanding view of the Pacific Ocean!

The High Dam at Aswan, Egypt is an example of poor land-
use planning. The dam was meant to provide water for crop
irrigation, but two serious mistakes turned the project into a
disaster. A 50% mistake in the original calculations for the evap-
oration loss did not account for 4 billion cubic meters of water
per year that are lost to the region's hot, dry winds, and an extensive system of underground cavities drains 12 billion cubic meters
out of the reservoir behind the dam. These losses could have been
predicted with field study, more careful calculations, and the design of a series of smaller dams (rather than one large structure).
Today, the lake behind the dam stands only half full, and the
level is not expected to rise for nearly a century. Other difficulties
caused by the poor planning of this dam include rapid silting
of the reservoir, decreased productivity of plants and animals in
the eastern Mediterranean, degraded soil quality, and increases
in malaria and other diseases. A little more time and effort in the
planning stages could have prevented such problems.

Land use must be evaluated in light of the geologic risks and
natural hazards. As the world population increases, more land-use

planning and site analyses will be needed for safe and effective use of available space. Thus, the need for consulting companies and engineering firms will increase also.

Job Duties

The job of a geologist working for an engineering or consulting company covers a wide variety of tasks. These generally include evaluation of specific sites, usually with an eye toward whether they are suitable for the proposed type of construction: dams, bridges, highways, power plants, housing developments, canals, tunnels, subways, or harbors. The evaluation of rock and soil material for its use in construction may also be a part of the job. In the process of these evaluations, assessment must be made of the natural hazards—landslides, earthquake potential, and rates of coastal erosion are among these. Engineering and environmental geologists deal most directly with geologic problems as they affect land use, including public safety.

The engineering geologist conducts site evaluation with the aid of many different kinds of information. The basic information sources include published literature, which describes what other people have found in the area of interest (or in similar areas); field observations; trenching, in order to get a look beneath the surface; and exploratory drilling. Other sources may include geophysical surveys, laboratory testing of samples, aerial photography, and subsurface studies.

An important aspect of working for a consulting company is the necessity to meet deadlines. When time is short, this may involve a tradeoff between the amount of work that needs to be done and the time available. The ability to accept the constraints of limited time (and often budget, as well) and perform the best possible job within those limits is important to all earth scientists considering a career in consulting.

After data have been collected, the information can be presented in many different ways. Written reports and oral presentations are the traditional methods, but consulting earth scientists will also find themselves testifying in government hearings and speaking at professional meetings. When a project is particularly controversial—like a dam or a nuclear power plant—a consultant may also find himself being interviewed by television, radio, and newspaper reporters.

Julia Van Auker is a marine geologist who originally planned to become a physical therapist. During her junior year in college, she happened to take a geology course from an excellent and enthusiastic teacher. The excitement rubbed off on Julia, and she decided to find out more about geology and geophysics.

She now works for a marine engineering firm, and her job involves many different activities and people. She explains:

"In my job, I meet and deal with a variety of people, from company presidents to administrators to high-caliber scientists to attorneys. My professional peers are divers, biologists, mathematicians, coastal engineers, urban planners, and oceanographers.

"My work experiences have included marine surveys, report preparation and presentations to government agencies and oil companies, and participation in projects ranging from writing an initial proposal through collecting and interpreting data to producing the final deliverables. I have acted in the capacity of electronics technician, geodetic surveyor, cartographer, hydrographer, shipping agent, and logistics coordinator."

Julia Van Auker
Marine Geologist
Tetra Tech, Inc.

Career Preparation

To prepare for a career in engineering or environmental geology, you should understand both the fundamental concepts of geology and the other earth sciences and have a knowledge of the principles and techniques of civil engineering. Although a degree in geologic engineering or geology is advisable, you should also acquire a good background in mathematics, basic engineering, field-mapping, and remote sensing, with coursework in hydrology, geochemistry, and materials properties. A small college may not have all the desired courses, so classes should be chosen carefully. Some colleges and universities offer majors in engineering geology or geotechnical engineering. Many smaller schools offer cooperative programs with nearby universities that have larger engineering programs; a student may spend a year or more taking courses at another university to strengthen his or her technical background.

The job outlook in consulting and engineering geology is best for graduates of geoscience departments that have a history of sending their graduates to engineering firms. If you are considering this direction within earth science, check the "track record" of the colleges or universities you are thinking about. A school with few contacts in industry will not be much help to you when you are ready to find a job. This is worth remembering, whatever field you plan to go into: when you are considering a school—undergraduate or graduate—ask about what type of jobs their graduates have obtained. Then compare this with the type of employment that you will be interested in. If the school already has a network of contacts in the field you hope to pursue, you will avoid some extra work when you start looking for a job.

Employment Outlook

Starting salaries for engineering geologists working for consulting companies vary with the employer and the person's education. In 1977, the median starting salary for engineering geologists (for all levels of education and types of employer) was about $15,000/year; this has increased steadily as the need for this type of consultant has grown. The employment outlook for this field is very good, reflecting the increasing concern for conservation of the earth's diminishing resources and the need for careful site planning of complex engineering projects. Conflicts between the growing pressures of society and the environment can be resolved by the information that can be provided by the consulting geosci-

An oceanographer retrieves a bathythermograph, a device that measures the variation of water temperature with depth. (Photo courtesy of the National Oceanic and Atmospheric Administration.)

TABLE IV

Typical 1982 Salary Levels for Geotechnical Consultants in the U.S.

Years Experience in Industry	Minimum	Maximum
Entry*–2	$16,900	$25,300
2–5	22,100	33,100
5–10	28,800	43,200
10+	36,700	55,000
Management	50,000+	

* Masters degree usually required.

Source: Major worldwide consulting firm (in Gaea, _1982, v. 5(1), p. 5)._

entists, who are most directly involved in the land-use issue. For those scientists qualified to help find solutions to these problems, employment opportunities will be great. Table IV shows typical salaries for geotechnical consultants in 1982.

The best opportunities in the past few years have been in consulting firms that work with practical applications of geologic principles. In these companies, some experience, or at least a solid understanding of engineering principles, is a big asset. Employers repeatedly emphasize the continuing need for hard-working, responsible entry-level people who are willing to do substantial field work and change projects frequently.

Consulting and engineering companies that work with environmental geology should experience a boom in the next few years. The disposal of radioactive, chemical, and solid wastes is a growing problem, and geologists with the expertise in locating and preparing waste disposal sites will be in great demand.

Some experienced engineering-oriented earth scientists are now being lured away to energy companies, which offer higher salaries and wider opportunities. A few others find jobs in government agencies. However, most of the openings for earth scientists interested in environmental science and engineering are in industry.

A field that is expected to grow significantly in consulting companies is meteorology. James E. Caskey, director of publications for the American Meteorological Society, has said that, in meteorology, "the real opportunity lies outside [government work] in private or consulting work for those who have entrepreneurial flair" (*New York Times,* 1978). Numerous industries are just now learning the advantages of having their own consulting meteorologists. For example, trucking firms are finding that they can ship goods more efficiently if, forewarned about the weather, their drivers can avoid storms or hazardous road conditions.

Many companies now receive a substantial portion of their funding from government sources. In an article in the *Houston Geological Society Bulletin* ("Future Trends in Professional Geology," December 1980), geologist Daniel N. Miller discussed this increasing control by the federal government ("federalism") as a possible trend in the beginning of the next century. He offered some hope for the survival of the private practice of professional geology, in spite of this trend, but this would depend on the expertise of the individual geologists. A solid background in either economic geology (minerals, metals, or fuels) or process-oriented geology (related to land use and natural hazards) should be the best insurance against unemployment or loss of funding should funding become increasingly dependent on sources within the federal government.

Some Points to Consider

An earth science career in industry doing consulting or engineering geology may be for you if

- You like a variety of tasks, including field work and presenting reports, both orally and in writing.
- You can deal with ethical questions of land use in potentially hazardous or environmentally sensitive areas.
- You like working with clients, presenting technical information to them in a way they will understand.
- You like to travel and would consider foreign assignments.
- You could face the prospect of moving every few years (many large engineering and consulting companies have offices throughout the country).

- You are prepared to meet deadlines on jobs, even when you know you could do a better job if you had more time.
- You are able to deal effectively with federal, state, and city governments.
- You are able to testify confidently as an expert witness in hearings or court cases.

Among the many disciplines of geology, engineering geology has always been the field most attuned to national direction; its practitioners are responsible for providing the most-needed early observations and recommendations for most construction and development projects, in predicting the nature and risk functions of natural hazards, and in providing methods for reducing negative impacts in and on the environment.

Allen W. Hatheway
Haley and Aldrich, Inc.
(Cambridge, MA)

"Engineering Geology,"
in *Geotimes*
(1981, v. 26, no. 2, p. 23)

CHAPTER 5

Careers in Industry: Minerals and Metals

Having now refuted the opinions of others, I must explain what it really is from which metals are produced.
—Agricola (1546)

Metals are necessary to most of the activities of civilization, including some that we never think about but that are crucial to our daily lives. Iron, glass, and salt are all commonplace earth materials that play important roles in human affairs. Some more exotic minerals are equally important; alloying of certain types of steel requires tungsten, chromium, tantalum, vanadium, and niobium.

The geographic distribution of metals is politically significant; countries with major deposits wield power because of their mineral wealth. For example, Rhodesia and South Africa have over 95% of the identified chromite in the world. Rhodesia alone controls approximately 85% of the world's high-chromium chromite. About half of all the chromite in the world goes into the production of stainless steel, and it is important in oil refining. Chromium is now a critical element for today's technology, and the United States currently imports over 90% of the chromium it uses. About 20 minerals, including chromium, cobalt, and manganese, are classified as "strategic minerals" because of their importance in defense and industrial production.

With the rising price of metals and other industrial minerals, exploration and exploitation have increased greatly. Mining companies often hire only a limited number of exploration geologists,

56

but there are good opportunities on the production end, working for well services and drilling companies, studying geologic samples, and interpreting laboratory data. These positions are generally available to college graduates who have not had additional experience or education. Companies often prefer to train their employees "in house," so your lack of experience is not a problem. A job in mineral extraction may have no special education requirement, and the pay is reasonably good. Annual salaries for mineral miners in Wyoming averaged around $24,000 in 1981.

Most major mining companies have separate exploration companies whose primary duty is to locate new economic deposits. Many of the large oil companies also have branches for minerals exploration. Throughout the western United States, the railroads have mineral rights (the legal right to extract minerals from below the surface) for the land they acquired to lay tracks; some of these companies also have mineral-resource divisions.

A recent addition to the range of job opportunities in minerals and metals is small-scale or individual prospecting. The national economy has made precious metal exploration practical for an individual or small company. As a result, numerous entrepreneurs are setting out on their own or establishing their own exploration, development, or production companies. This type of employment tends to be a high-risk venture, but some people find it an exciting gamble. If a project is successful, the financial rewards can be great (imagine finding your own gold or silver mine!).

Mineral and metal exploration covers the spectrum from gold and silver (and diamonds) to sand and gravel. In fact, sand and gravel, along with other construction materials, are important economic resources. For example, Trinity County, California, which was a focal point for California gold mining in the 1850s and still produces gold today, yields about 50% of its mineral production today in sand and gravel—and most of the remainder is other construction materials, including stone and pumice. We all use roads and buildings constructed with sand and gravel, but we seldom think of these everyday materials as mineral resources. In the mid-1970s, the amount of sand and gravel used in the United States averaged 4.1 metric tons for each person—compared to only 3.8 metric tons of petroleum per person!

The economics of precious minerals have also renewed in-

Research geologist surveying a limestone quarry in the vicinity of Weeping Water, Nebraska. (Photo courtesy of Jay Fussell, University of Nebraska.)

terest in diamonds. Several kimberlite pipes have been studied recently by the Wyoming Geological Survey, and scientists at several universities have been trying to understand the relationships between locations where diamonds are found and the theory of plate tectonics. A few years ago, an economic geologist might have been considered unrealistic to be concerned with diamonds, but the world economy now makes it a very reasonable interest.

Career Preparation

The willingness to do support services on the production end of the minerals industry is often a foot in the door toward a research position. Many exploration geologists got their start by "sitting on wells"; this involves working at a drilling site, sampling the rocks that are being cored and the rock fragments coming from the hole. An additional advantage to such entry-level jobs is that they are usually available to college graduates with no additional experience. Other entry-level jobs include logging wells, placing geophones (receivers that detect energy reflected from buried geological structures), washing microfossil samples in paleontology laboratories, and working as a field assistant.

If you are interested in a career in minerals and metals, you should try to acquire a broad background in the earth sciences. New exploration techniques involve information that may not be traditionally geologic. Exploration for copper, for example, takes advantage of some of these new approaches. An area with a copper deposit will have soils with high copper concentration, up to several tenths of a percent. If a plant growing on that soil has a low copper tolerance, it will die. With this information, earth scientists have used satellite photographs of Norway, which show large areas of yellowed, copper-poisoned vegetation, to discover copper-rich soils. Other plants are able to concentrate copper in their tissues. One study of an African mint *(Aeolanthus biformifolius)* in Zaire revealed as much as 1.3% copper in the plants— and the average copper content (grade) in ores that are mined in the United States today is less than 0.6%! Studying this concentration of elements in plant tissue is called "geobotanical prospecting." Clearly, a diverse educational background is useful in this sort of work.

Many companies involved in mineral exploration and development are small, and each employee is responsible for a variety of duties. Experience or knowledge in economics and business administration can be extremely useful in both smaller companies and larger corporations.

Much mineral exploration is controlled by the Bureau of Land Management (Department of the Interior), and complex regulations may restrict the exploration and development of a land area. Some understanding of land-use laws and legal matters can make an individual very important to the company. In fact, this is one employment area in which joint degrees in geology and law may be a real advantage.

My interest in geology began during my freshman year in college. The courses that I was taking seemed far removed from reality, except one . . . Introductory Geology. The professor who taught the second semester of Intro was very absorbed with and excited about geology. His course material reflected his involvement. I remember a particular portion of the course that involved the practical application of geological methods and concepts. We students were asked to determine the best locations for various man-made projects such as a sanitary landfill, a huge factory, etc. In short, geology suddenly seemed important, relevant, dynamic. . . .

Today, for me, geology remains an interesting and rewarding profession. I consider myself fortunate to have found a career that is both enjoyable and lucrative.

David P. Desenberg
Senior Exploration Geologist
TXO Minerals/Texas Oil and Gas

Employment Outlook

Minerals and mining have been described as "never brighter" as a potential source of employment. This is particularly true of jobs in industrial minerals, which are growing faster than jobs in the metals industry. In fact, minerals and mining were rated #1 in outlook in the October 1980 *New York Times* recruitment survey.

Mineral fertilizers are slowly but steadily increasing in production. In 1979, phosphates and potash production increased 7% and 3%, respectively. These commonplace minerals are not exempt from the international politics that affect chromium and gold. For example, President Carter banned the shipment of phosphate fertilizers to the Soviet Union in February 1980 to protest their intervention in Afghanistan's national affairs. These

Geologists explore below the earth's surface as well as above it. These geologists are preparing to explore a limestone cave system in Pennsylvania. (Photo by the author.)

shipments were valued at $97 million in 1979, and they had been expected to double in 1980.

Paralleling a retreat in the nuclear energy industry in the wake of Three Mile Island, there has been a cutback in uranium exploration and production. Many uranium geologists have found employment in countries where uranium exploration is increasing (like Australia), and others have transferred their skills to petroleum geology. Being able to anticipate trends is an important attribute in this industry; the time required to construct a new nuclear power facility is three to five years, and it may take 15 years to locate and develop the necessary new mineral deposits. This long lead time demands foresight—and possibly a crystal ball—for decision makers in the minerals industry.

The average salary offer to an entry-level employee in March 1980 was $18,000 for a graduate with a bachelor's degree and $24,000 with a master's degree. The best salaries are found with the minerals groups within petroleum companies. Excellent opportunities can also be found with smaller exploration companies. Entrepreneurs, both individuals and companies, are often involved in speculative, high-risk ventures, but successful results can be extremely rewarding. Demand for mining geologists is especially high in Alaska.

Almost 8,000 people with degrees in mining engineering were employed in 1980; most of these worked within the mining industry but others worked for colleges and universities, government agencies, and private consultants. This field is expected to continue to grow beyond the 1980s. Particularly important areas are mining safety, energy resources, secondary recovery of minerals, and extraction of resources from the ocean floor.

Some Points to Consider

A career in the minerals and metals industry might be right for you if

- You would like working in undeveloped parts of the country.
- You like to travel.

- You like the excitement and uncertainty of exploration.
- You are interested in safer and less environmentally harmful ways to extract resources from the earth.

Realistically, everything that we have comes from the earth. All the materials we use are ultimately derived from rocks—even plants depend on the soil, and the atmosphere was produced from rocks. The writer John McPhee recognized this dependence, as shown in the following passage. This excerpt is part of a discussion about the conflict between environmental conservation and environmental utilization.

"Most people don't think about pigments in paint. Most white paint pigment now is titanium. Red is hematite. Black is often magnetite. There's chrome yellow, molybdenum orange. Metallic paints are a little more permanent. The pigments come from rocks in the ground. Dave's electrical system is copper, probably from Bingham Canyon. He couldn't turn on a light or make ice without it. The nails that hold the place together come from the Mesabi Range. His downspouts are covered with zinc that was probably taken out of the ground in Canada. The tungsten in his light bulbs may have been mined in Bishop, California. The chrome on his refrigerator door probably came from Rhodesia or Turkey. His television set almost certainly contains cobalt from the Congo. He uses aluminum from Jamaica, maybe Surinam; silver from Mexico or Peru; tin—it's still in tin cans—from Bolivia, Malaya, Nigeria. People seldom stop to think that all these things—planes in the air, cars on the road, Sierra Cups—once, somewhere, were rock. Our whole economy—our way of doing things, most of what we have, even our culture—rests on these things. Oh, gad! I haven't even mentioned minerals like manganese and sulphur. You won't make steel without them. You can't make PAPER without sulphur. By a country's use of sulphuric acid you can almost

measure its industrial capacity. The top of Mount
Adams has been prospected by sulphur companies.
Did you know that?"

John McPhee, 1971
Encounters with the Archdruid, pp. 48–49
Farrar, Strauss and Giroux, New York.

CHAPTER 6

Careers in Government: Federal, State, and Local

Once a photograph of the Earth, taken from the outside, is available, once the sheer isolation of the Earth becomes plain, a new idea as powerful as any in history will be let loose.

—Fred Hoyle (1948)

Federal Government

Nearly 25% of all earth scientists in the United States are employed by the federal government. The U.S. Geological Survey (established in 1879) is the largest federal employer of geologists, providing jobs for more than 1,600 men and women. Other government agencies that employ geologists are the Bureau of Reclamation, Bureau of Mines, Office of Surface Mining, Bureau of Outdoor Recreation, the National Park Service, and the Forest Service. The Soil Conservation Service (U.S. Department of Agriculture) employs 1,200 soil scientists, and the Agricultural Research Service also employs earth scientists. Employment for oceanographers, meteorologists, and geologists can also be found with NOAA (National Oceanic and Atmospheric Administration), the Army Corps of Engineers, and NASA (National Aeronautics and Space Administration).

The usefulness to the government of information on natural resources has been known for many years. Indeed, the government clearly recognized that need when Lewis and Clark were requested to collect geologic information during their travels in 1804–1806.

National Park naturalists have many jobs, including detailed studies of the park and conducting nature walks for visitors. (Photo courtesy of the National Park Service.)

Other western surveys, like the 40th Parallel Survey (begun in 1867), were almost exclusively intended to collect geological information. Governments of countries that are only now becoming aware of their resources, particularly African nations, are actively recruiting geologists and hydrologists to help locate important materials. Thus, a career in government geology need not be limited to the United States.

Earth scientists have continued to be pioneers and explorers. Geologist Harrison Schmitt collected rock samples while on the moon, and several other earth scientists, including a woman, are working in the astronaut program now.

All of the Apollo astronauts were given considerable geologic training in preparation for "field work" on the lunar surface.

Craters are the most abundant landforms in the solar system, but erosion of the earth's surface erases such features rapidly. Thus, the astronauts' training included a field trip to Meteor Crater in Arizona, one of the best (and most recent) examples of a meteorite impact on Earth. Much of this part of the astronauts' training was handled in conjunction with the U.S. Geological Survey's Astrogeology Branch.

After the success of the first Space Shuttle flight, NASA began further recruiting for the astronaut training program (address in appendix). Experience commanding high-performance jet aircraft is still required for astronaut pilots, but no flight training is necessary for mission specialists. The best opportunities for earth scientists to go into space will be in this category; mission specialists are responsible for the scientific operations on the spacecraft, including experiments and instruments. Activities during a mission may involve taking photographs of the earth's surface, "walking in space" to inspect or repair equipment, or performing experiments with materials under zero-gravity conditions.

Job Duties

Projects with the federal government may include field mapping, resource evaluation, geochemical water studies, oil and gas resource evaluation, and leasing and conservation studies. Employees of the National Park Service and Forest Service may conduct tours or oversee and maintain national monuments or parks. The administrative duties may also include overseeing mining or drilling operations within their boundaries.

Barbara Faulkner began college believing that she wanted to be an English major, but an introductory geology course changed her mind. After completing her B.S. in geology, she spent a season working for the National Park Service. Since then, she has returned to graduate school and earned a master's degree, specializing in paleontology. She describes her job:

"As a National Park Ranger, I lived and worked on Cumberland Island National Seashore, a semitropical, barrier island on the Georgia coast. I had an entry-level position (government service-level 4) and worked as an 'interpreter' of the human and natural history of the island.

"As a Park Interpreter, I had the challenge of developing and conducting natural history programs in the field. My programs included beach walks, salt marsh tramps, bird-watching trips, and cross-island hikes. Most of these included far more biology than geology. During the beach walks, however, the visitors and I discussed tides, waves, currents, sediment transport, beach profiles, barrier island migration, and even theories on barrier island formation; all of this was in addition to discussions of the life habits of the shells we found, the adaptation of the dune plants, and other biological topics. A geologist's perspective can really stimulate the general public, even on an island made of sand and mud.

"Some of my additional duties were quite exciting, such as flying in a helicopter over the island to locate egret rookeries or joining a fire crew to fight a forest fire that was raging out of control. Other duties were downright boring, such as vacuuming the rug in the Visitor's Center every morning. I occasionally worked by myself; I spent some days taking photographs for the park files, identifying plant specimens, or hiking the trail system. Most days, however, I interacted with the general public. During my five-month season, I spent countless hours greeting visitors, giving orientation talks to backpackers, and answering questions from almost anyone. But the core of my job was designing and giving natural history programs, and this remained my favorite aspect of the position."

Barbara L. Faulkner
National Park Ranger

The federal government is supposed to "serve the people," and so jobs with the government usually involve interaction with the public. This may occur in public meetings, congressional hearings, or discussions with landowners. An essential quality for a federal employee is writing ability; most U.S.G.S. research, for example, is published, and therefore all professional employees spend a considerable amount of their time writing their research results.

Employment Outlook

Employment opportunities with the U.S. Geological Survey and the other federal agencies employing earth scientists are currently quite restricted, reflecting hiring freezes and budget cutbacks throughout the federal government. Jobs are still available, partly because federal employees continue to take jobs outside of the government or retire or because federal funding increases for some programs, but competition is severe for the openings. Many new employees have to accept part-time or temporary work, sometimes for several years, before they are able to move into a fulltime position. In 1979, the average annual salary for geologists working for the federal government was over $27,000.

Despite the movement toward economy in the federal government, there is at least one area within the U.S.G.S. that may be expanding: the Conservation Division. This branch oversees the leasing and regulation of oil and gas production, and with the increasing importance of energy, this division is expected to hire a substantial number of new geologists and geophysicists. As long as the conservation division continues to perform this function, geophysicists of all levels will be needed; geologists who have advanced training or work experience with minerals and energy research or production will also be important. Work in the conservation division includes resource evaluation, development, and production of energy-related materials, such as oil, natural gas, oil shale, coal, minerals, and geothermal steam. Geophysicists will be needed especially for offshore work. Locations of most job openings will be in Albuquerque, New Mexico; Casper, Wyoming;

Metairie, Louisiana; and Anchorage, Alaska; as well as the major U.S.G.S. centers in Denver, Los Angeles, Menlo Park, California, and the national headquarters in Reston, Virginia, near Washington, DC.

"Earth science uses many researchers to search for solutions to an immediate practical problem, such as securing minerals from the earth. However, many thousands of scientists carry out the other type of research, which we'll call basic research. Those who do this work seek only to find new physical principles, discover Mother Nature's secrets and how She operates. An immediate practical benefit is not necessarily apparent in such research efforts. And, a practical benefit may never materialize. On the other hand, with basic research you never know. After all, as the classic saying goes, of what use is a newborn babe?

"My own research includes a little of both types of research but definitely is closer to basic than to applied research. Within the federal government, the people who make decisions on funding my work are rarely idealistic research scientists. They and their administrative superiors generally must justify their decisions on the allocation of funds. My work therefore must fit within the overall goals of the U.S. Geological Survey and its divisions. Fortunately this is not a problem, as the goals are necessarily broad, and I try to choose topics that I hope have the potential of eventually proving useful.

"One of the unpublicized rewards of doing research in the earth sciences is the fabulous people you meet. Sitting in a farmer's kitchen enjoying a chat with him and his wife; talking with the truck driver who delivered the supplies and materials for an apparatus I had to construct; touring a pasture with a local historian who showed me the nearly invisible remains of an old bridge and the channel of a former mighty river; discussing with a photographer the best way to prepare and display some color slides; feeling the enthusiasm rise during a discussion of new research findings and ideas with colleagues at symposia; these and many other scenes bring fond memories to mind as I reflect on my own past research projects.

"An immediate reward of research in earth sciences is the uplifting of the spirit and the general pleasure obtained from seeing the pristine out of doors. Whether your trail leads to rivers (as mine usually does), or to prairies, mountains, deserts, oceans, glaciers, or the atmosphere, you can't beat the scenery."

Dr. Garnett P. Williams
Hydrologist
U.S. Geological Survey

The Geologic Division is also expected to experience some growth in the next few years. The emphasis here is also on energy and mineral resource evaluation. The Menlo Park (San Francisco area, California) and Anchorage (Alaska) offices have the greatest promise for expansion. The Alaska Division is gradually being moved from Menlo Park to Anchorage, and for geologists who are willing to move to the far north, opportunities are opening slowly but steadily.

Areas of research in which the U.S.G.S. is expected to grow in the next decade include energy and mineral resource evaluation (as already mentioned for both the conservation and geologic divisions), remote sensing, urban and environmental geology, natural hazards, seismic exploration, continental shelf drilling, water resources, climatic change (especially related to food and energy production), geothermal heat flow, and ocean mining.

Historically, the U.S.G.S. employs many women; indeed, it is an important source of employment for women in geology. Approximately one third of all the full-time and nearly half of the part-time U.S.G.S. employees are women. This includes technicians and technical support staff, as well as geologists. This employment ratio is expected to continue at about the same percentage for the foreseeable future.

Soil scientists mapping soil types onto an aerial photograph in the Washington, DC area. (Photo courtesy of
U.S.D.A.—Soil Conservation Service.)

Sometimes predictions concerning the expected trends in
earth science can be altered drastically. A recent example of this
is the eruption of Mount St. Helens, which created $11.5 million
in funding and 18 new jobs for the U.S.G.S. Intensive research
will be going on in the Mount St. Helens area for many years
to come. In the same way, a major earthquake would generate
considerable activity and job opportunities for earth scientists.
Earth science is a dynamic field, and the employment opportunities are continually changing.

Some Points to Consider

A job with the federal government may be right for you if

• You like the idea of collecting objective data on critical
  national problems.
• You like interacting with all fields in earth science.

- You want to work in a collegial atmosphere.
- You can live with the uncertainties of salary increases and retirement programs that are associated with any job in the federal government.
- You wouldn't mind living in one of the major urban centers where government offices are located (Washington, DC, San Francisco, Denver)—or in one of the outposts, like Casper, Wyoming, or Metairie, Louisiana.
- You like helping and interacting with the public.
- You are good at expressing yourself both in speaking and writing.

State and Local Government

State and local government agencies produce geologic and soils maps, evaluate resources, and interact with residents as a general public information service. State employees are often called on to testify as expert witnesses in legal cases or before legislative committees. Local agencies employ earth scientists to consult on problems unique to the area, including construction, agriculture, weather patterns, natural hazards, and economic resources.

This field has expanded greatly within the last five years. Some state surveys are essentially "one-man shops," while others employ hundreds of people. The larger state geological surveys are generally located in states with significant mineral or petroleum resources. In mid-1980, 1,169 geologists were employed by state surveys (77% full time); 15% of this total were women. These positions are not always labeled "earth scientist" or "geologist"; they may be described as "resource-use manager," "inspector," or any one of a number of other job titles.

Early state surveys were established to assess the mineral and economic resources of that state; many early surveys included biologic studies. North Carolina established a state survey in 1825, followed by Massachusetts (1830) and Maryland (1834). Virginia organized its survey in 1835—the same year that Great Britain established its first national geological survey. Today, the orientation of state surveys is more toward land use, and their major role is providing information for state and local decision makers as they plan the future of the area.

Other state agencies besides geological surveys or natural resources departments employ earth scientists, including highway departments, public health offices, planning and development commissions, regional planning boards, divisions of mines, water boards, and realtor boards (which are especially interested in engineering geologists). In many states, city and county governments also employ earth scientists; such positions include planning departments, construction inspectors, and zoning commissions.

State geological surveys produce a variety of publications, which are available free or at low cost to state residents. These may include detailed "quadrangle" reports, covering areas that are 7.5′ or 15′ of latitude and longitude across, information about mineral or fossil collecting in the state, and reports on local mining history. Many states also publish magazines that are available free or by subscription. Examples of these include the monthly *California Geology* and the quarterly *Pennsylvania Geology*. Information about these publications can be obtained by writing directly to the state geological surveys (addresses in appendix).

"Ultimately, the prime clients of a state geological survey are the citizens of the state, and that is why their needs should be served first and above all else. If that is done and done well, requests for information and publications will come in from all over the nation. The state is the place where service begins but never the place where it ends. Serving geological needs on the state level means that the state provides perspective, or an angle of vision, from which to begin one's service. But like ripples that spread out in all directions, such efforts are never confined to the boundaries of the state."

Jay Fussell
Publications Officer and Associate Professor,
Conservation and Survey Division,
Nebraska Geological Survey,
University of Nebraska-Lincoln

Career Preparation

Geoscientists working for a state or local government often have to cover a wide range of expertise. In dealing with the public, they have to handle questions ranging from "What is this rock?" to "Will the river flood my property this spring?" If you are interested in this type of employment, you should obtain as broad an educational background as possible.

Thornton L. Neathery, of the Geological Survey of Alabama, surveyed the state geological surveys in 1980. He found that the five geology courses that state surveys are most interested in seeing on an applicant's transcript or resume are structural geology, field geology, petrology/petrography, stratigraphy, and geohydrology. Other important coursework includes other physical sciences (chemistry, physics, biology, meteorology, and oceanography), calculus, English composition, computer science, and economics.

For a position with a future, advanced degrees are usually required. Graduate degrees are viewed by employers as evidence that the applicant has the ability to conceive a research problem, pursue it, and carry the entire project to completion. In fact, many state surveys prefer a master's degree over a PhD, because people who have earned doctorates sometimes have narrowed their interests too much to be useful for the general work required in state surveys.

The most important attributes for someone working for a state or local government are enthusiasm, experience finding and using reference sources, the ability to start and complete a project, and the ability to work with the public. Because these state offices are generally small, supervision may be minimal, and therefore it is particularly important for an employee to be able to work on his or her own.

Employment Outlook

Jobs with state and local governments become available because either (1) a current staff member retires or resigns or (2) the basic programs are changed so that new positions are created.

Most of the positions for earth scientists with these agencies are not advertised widely, and thus it is often necessary to "beat the bushes" to find these jobs. Many require several years of experience, and advanced degrees (master's level) are preferred.

For those who are qualified, good pay is available. With a bachelor's degree, average salaries nationwide are in the $15,000–19,500 range. With a master's degree, this increases to $17,000–22,500, and with a PhD, salaries average between $21,500 and $27,500.

With cutbacks in federal budgets, we can expect to see more cooperative work between the federal government and state and local agencies, rather than projects funded exclusively by the federal government. Cooperation between governments and the private sector will probably also increase.

Some Points to Consider

A career in earth science with a state or local government may be right for you if

- You like diverse tasks.
- You are good at public interaction.
- You like the idea of saving money for a region or a community—and perhaps even saving lives.
- You would like to develop local roots and feel part of a community.
- You are looking for job security.
- You won't object to long hours and managing with limited resources (this will often require some ingenuity on your part).
- You're not afraid of a job that has high stress levels and sometimes has demanding deadlines.

CHAPTER 7

Careers in Teaching and Research

"Why," said the Dodo, "the best way to explain it is to do it."

—Lewis Carroll (Charles Dodgson)
Alice in Wonderland,
Chapter 3

Teaching

The one area in the earth sciences that is *not* expected to boom in the next decade is teaching. Educators are the role models that students are confronted with every day, and it is natural that this should seem appealing when students are planning careers. Indeed, teaching has many advantages, but it is not without its problems. The only optimistic aspect of geological and earth science teaching is that the outlook is not as dismal as it is in some other disciplines.

Most of the earth scientists teaching in colleges and universities are geologists and geophysicists (Fig. 7.1). Oceanographers, meteorologists, and soil scientists also hold faculty positions, but the majority of these scientists work for government agencies and private industry.

Opportunities for teaching earth science in grade schools or high schools vary according to the locality. Public schools almost always require a state teaching certificate, although these are not required to teach in a private school; if you are interested in teaching in a grade school or high school, check on the regula-

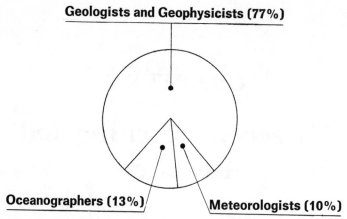

Fig. 7.1 Earth scientists employed by colleges and universities as of 1978 (total = 13,500). Percentage of soil scientists is negligible. Data from the *Occupational Outlook Handbook,* 1980–1981, U.S. Bureau of Labor Statistics.

tions concerning teaching certificates in the state or city you would like to work in. The prospects may be good in a specific region or town, but the nationwide trend is toward a decreasing emphasis on earth science education at the secondary school level. A renewed national interest in the quality of science education could —and, we may hope, will—reverse this trend.

One of the major factors contributing to the decrease in college faculty hiring is the decreasing number of 18-year-olds entering college, related to the falling-off of "baby boom" enrollments. This effect will not be felt uniformly. It should be worst in the Northeast and Midwest and less severe in the Sun Belt. The enrollment decreases will create problems for some of the more faculty-intensive programs within departments, especially field-geology courses.

The tenure crunch is also a problem. Colleges hired many new PhDs in the 1960s, and these teaching positions will remain filled for the next 15–25 years. Many universities are consequently overstaffed, with faculties that are not yet at the retirement age. In fact, some states are now passing laws that forbid mandatory retirement at any age (Rhode Island and Utah are examples of this), and this will further complicate the employment problem.

Two-year community (junior) colleges are having many of the same problems as their four-year counterparts, but employment opportunities in this area seem to be somewhat better. Teaching at a two-year college generally requires a master's degree, and some secondary school teaching experience is useful. These teaching positions sometimes prove to be lonely outposts; although some community colleges have excellent, well-staffed

Geologist observing lava flows on the flanks of Kilauea Volcano, Hawaii. (Photo courtesy of the U.S. Geological Survey.)

geology departments, most have only one or two members. Although the amount of professional interaction may be limited, emphasis on teaching (rather than research) offers broad opportunities for innovative courses.

Many colleges and universities already have programs in earth science education, combining geoscience courses with the required education classes. This type of program can be very efficient, saving the student the trouble of setting up a self-developed major . . . or spending an extra semester student teaching to earn a teaching certificate.

One advantage in teaching is the ability to change from academics into industry. Geology has more opportunities than almost any other academic field for changing to a job in industry. Many teachers, including some who are nationally known for their teaching ability, are being lured away by the significantly higher salary offers in the petroleum industry or with engineering and consulting firms. This "brain drain" especially affects geophysicists, hydrologists, and geologists specializing in sedimentary rocks (where oil and gas are found), and this trend is expected to continue. Those scientists who are concerned about the future of science and technology in this country are especially troubled by this situation: If the good teachers are leaving the profession, who will train the next generation of scientists? Indeed, this question should concern us all. Good earth scientists who choose to become teachers, in spite of higher salary offers from industry, will be important in providing future generations of qualified scientists.

"I have taught for 12 years in a small liberal arts college and continue to enjoy it. Compared to teaching in a large university, teaching in a liberal arts college requires one to be a 'jack of all trades,' but at the same time one must be a master in one's own specialty field. The majority of my teaching is traditional—introductory geology, field geology, structural geology, geomorphology. But even within geology, the courses I teach would be covered by two or three people at larger schools. I have also had the opportunity to teach interdisciplinary courses with a historian on the Lewis and Clark Expedition, with chemists, physicists, and biologists in Environmental Science, and a freshman seminar in which the readings ranged from plate tectonics to Plato, Bacon, Tillich, Skinner, and Whitehead.

"Small schools pride themselves on individual attention for students, and much of one's time is devoted to these things. Geology comes first, but the other opportunities are here for those who wish to work with colleagues in other disciplines. One's research often must take a back seat to the immediate needs of students during the school year. Time for intensive research must be found during vacations, in the summer, and during sabbatical leaves. However, I do pursue research with upperclass students working on independent projects during the academic year.

"Monetary rewards are less substantial in the academic setting than in industry for geologists, but there are less tangible rewards. One is that I am largely my own boss and may do research on what I wish and can set my own work schedule, aside from the necessity to meet classes at prescribed times. If I prefer to work at odd hours—early mornings, evenings, weekends—I may do it. There is also the reward of seeing relatively unsophisticated students become quite sophisticated as geologists, and of hearing that former students are doing well after they leave here.

"I enjoy working with a diverse group of people. These include not only geology majors, but non-majors as well. These contacts occur as often outside the classroom as in it. One must deal with colleagues in a diversity of departments, with deans, counselors, and financial officers of the college. Encounters with specialists in my own field of specialty occur primarily at professional meetings, which I attend once or twice a year.

"The best preparation for a career as a teacher in a small liberal arts college is to attend a school with a good geology department. I would recommend a geology department in a larger university with a graduate school where there is a larger staff, library, and more sophisticated equipment. I think that a firm foundation in geology at the undergraduate level is important, and this should be supported by courses in mathematics and the allied sciences. The geology department should have a good balance between field and laboratory studies."

Dr. Noel Potter, Jr.
Geology Professor
Dickinson College

Research

College and university teaching usually requires independent research in addition to teaching responsibilities. In fact, estimates of the percentage of basic research done by universities in this country, rather than private industry, range from two thirds to as much as three fourths. Although the educational institutions may sometimes support the research financially, faculty members are usually required to obtain funding for their research projects outside of their college or university. This outside funding may take the form of government grants, contributions from private foundations, or support from businesses or industries that are interested in that particular research problem. Major supporters of geological and earth science research have included the Geological Society of America, NASA, the National Science Foundation, the U.S. Geological Survey, the Departments of Defense and Energy, and several major oil companies. Federal support of university research reached a peak in the late 1960s, but new national emphases on mineral resources may encourage renewed government support of university research. The government is shifting its interests from basic, "pure" research to applied research related to identified national needs. Funding from industry, including petroleum and steel companies, also appears to be growing.

Although it's usually an exciting challenge to combine both

teaching and research, some faculty members are frustrated by the hazy boundaries between the two, and the question of whether teaching or research is more important to their institution concerns many teachers. For those who find that they prefer the research over the daily deadlines of meeting classes, there are several possibilities that allow independent research. Some private research companies provide the facilities for basic research without any teaching responsibilities. These laboratories often operate very similarly to a college or university science department, allowing geologists and earth scientists the opportunity to devote themselves full time to the research problems at hand.

Many private research companies require their scientists to obtain their own funding, frequently from government agencies. Examples of these private research organizations include the Jet Propulsion Laboratory of the California Institute of Technology, Lawrence Livermore Labs, Battelle Memorial Labs, the Aerospace Corporation, and Sandia Labs. Several laboratories that were once run by the Atomic Energy Commission are now National Laboratories, which perform research under federal contracts. These include Oak Ridge (Tennessee), Argonne (Illinois), and Los Alamos (New Mexico). A growing percentage of the research at these laboratories is now concentrated on geological sciences and mineral resources.

A national laboratory such as Battelle Northwest is a focus for multidisciplinary sponsored research, and it can be a tremendously exciting place for a technically-oriented, adaptable person. I work in the area of disposal of radioactive wastes in mined geologic repositories [e.g., salt mines, deep mines in granite], primarily in the computer modeling of geologic and hydrologic processes which might affect the integrity of such a repository over a million years. This puts me at the forefront of state-of-the-art, interdisciplinary work in modeling geologic processes, with a very real and critical problem for immediate application. In addition to being a technical contributor, I am responsible for managing a large part of the program, which involves me directly in the pursuit of a scientific effort on a much larger scale and of greater variety than I could achieve as an individual. This experience could have been shattering had I not had a diverse background in engineering and geology, experience as an engineer in the Apollo and Viking programs, and a stint as an assistant professor to increase the breadth and depth of my abilities as an earth scientist.

The management of hazardous wastes is going to be an increasingly important national goal, and it will need the services of earth scientists with broad training in the more quantitative aspects of physical processes—hydrological, geological, biological, geochemical, and geotechnical. In addition to multidisciplinary training, a period of intense individual research in an academic setting (e.g., faculty or postdoctoral) is a great aid to making the sometimes hectic, rough-and-tumble pace of multidisciplinary, multi-sponsor research an exhilarating rather than unpleasant experience.

Dr. Michael G. Foley
Senior Research Scientist and
    AEGIS Associate Program Manager
Battelle, Pacific Northwest Laboratories

Employment Outlook

The job market for teaching at the junior college, college, and university levels is so tight that a PhD is now required for nearly all positions, but it does not guarantee a job. Many of the jobs that are available in teaching are either short-term "replacement" positions or part-time jobs, which do not usually have fringe benefits associated with them. One- or two-year appointments are also available at some schools.

If what you want to do most in the whole world is to teach, you should not be discouraged. The jobs do exist. Approximately

650 new geoscience faculty were employed between 1977 and 1980. They are not easy positions to obtain, however, and many of the rewards are more internal than monetary. Salary scales are generally low, but summers are usually free for outside consulting, research, field work, or writing. Many college professors supplement their income during this vacation time. One of the frustrating aspects of college-level teaching and its relatively low salaries is watching one's students graduate and receive salary offers from industry that are considerably higher than one's own salary!

As mentioned above, the current national deemphasis on science education will not help the employment situation for secondary teachers, but the effects of this trend will vary locally. Salaries will also be determined by city or county regulations. Although a bachelor's degree and a valid teaching certificate are the minimum requirements for teaching in a public secondary school, many school districts pay higher salaries if a teacher has additional coursework beyond a bachelor's degree.

Private research companies have been well supported by grants from the federal government in the past. The changing emphasis in the federal government from basic to applied research (and especially research that has military and defense applications) will affect many of these research laboratories. Some will shift their concentrated efforts to agree with those of the funding sources; some will find new sources of grant money; and some will find themselves unable to survive. When considering a job with a research company that is dependent upon outside funding, be sure to find out about its plans and expectations for the next few years.

Some Points to Consider

A career in teaching or research in the earth sciences may be for you if

- You like being your own boss.
- You are able to manage your own time without supervision.
- You are motivated enough to carry on your research by yourself.

- You feel that the free summers would compensate for the much lower salaries in teaching.
- You like helping people understand about earth science.
- You can deal with administrative details, like grant-writing, grading papers, organizing field trips, and serving on college administrative committees.
- You can handle the uncertainties of tenure decisions (for teaching) or having to find your own grant money (for research).

CHAPTER 8

Preparation and Education

There is something fascinating about science. One gets such wholesale returns of conjecture out of such a trifling investment of fact.

—Mark Twain (Samuel Clemens),
Life on the Mississippi (1883)

A fundamental requirement for a geologist is a good imagination and the ability to visualize things in three dimensions. Earth science is an intuitive science; being able to pull together a variety of types of information and understand how they fit is an essential talent. Being an accurate observer and the ability to distinguish between observations and interpretations are equally important.

The educational background needed for an earth science career is very broad, including courses in all the natural sciences (geology, physics, chemistry, biology), mathematics (at least through calculus), English and communications skills, and foreign languages (especially German, French, or Russian). Computer science and programming skills are useful as well, and they will probably be required soon.

Graduating college seniors have excellent job opportunities, especially in geology and geophysics, but a master's degree is usually a minimum requirement for any professional advancement. A doctorate is necessary in some areas, especially college-level teaching and professional research. In many academic and government jobs, pay scales are tied to the years of education completed, and so it becomes a monetary advantage to have additional education.

Opportunities for anyone without a college degree are extremely limited. Some openings are available for support person-nel, including technicians and staff on production teams (oil rigs, mines, etc.), but even advancement to a junior management posi-tion is unlikely to be available to people without a college education. Thus, anyone planning a career in earth sciences should also plan to obtain at least a bachelor's degree. There is no need, however, to wait until reaching college before you start planning your career.

High School

For anyone planning a career in science, the high school background is very important. This is where obtaining a strong foundation in mathematics, physics, biology, English, and chem-istry will help immeasurably in the years to come. Many college courses are taught with the assumption that the students have had a solid education in high school. Although remedial English and math are becoming increasingly common in colleges now, they are basically a repetition of material that should have been cov-ered in high school. (The moral of this is: Pay attention when you take the course for the first time, and you'll be able to go on to new topics when you get to college.) In some high schools, the student may have to insist on taking certain courses in order to obtain an adequate background. Unless a student expresses a real interest or aptitude in science, he or she may be discouraged from taking a curriculum with several science or math courses. One of the advantages of early planning for a career is being able to take classes in high school that will be helpful later.

High school is not too early to begin developing professional associations with earth scientists. If there is a college near where you live, try to visit it; you'll find the professors and students happy to talk with you. If you are genuinely interested, you may even manage to be included in an occasional field trip. Some of the students, particularly if it is a college (i.e., not a university with both graduate students and undergraduate students), may need an extra hand working on field projects, and by volunteering your help, you can pick up valuable experience. [Similarly, college students needing field assistants might consider science students from nearby high schools for field assistants.]

Geology students recording measurements and descriptions of an outcrop (Photo by J. Kruissink.)

College

There are over 450 academic intitutions in North America that award degrees in one or more branches of earth science. Available courses and institutional requirements vary, so you should try to find a school that has a program that corresponds with your interests and has students that have similar abilities to yours. An important source of information on the geology departments of colleges in the United States and Canada is the American Geological Institute's Directory of Geoscience Departments. This book is published yearly and is available from the AGI (address in appendix). It includes a listing of faculty members in each department, their interests, the degrees that the school offers, the number of graduate assistants, and the addresses and phone numbers of the departments. A list of some schools that offer bachelor's degrees in geology is also included in the appendix at the end of this book. Specific information can be obtained by writing to the department at the address shown in that list.

Selecting a college is an important decision deserving careful consideration. You can get a good education at any college or university; the three major points to evaluate in your decision about the best school for you are (1) size, (2) location, and (3) academic program.

Small colleges and large universities are fundamentally different; both have unique advantages. Small colleges offer personal attention. Class and laboratory enrollments are small, and most of your professors will know your name. The number and variety of courses offered will be limited by the size of the faculty in the department. Most small colleges do not have all the modern equipment they would like, but all the departmental equipment is usually available for undergraduates to use in their projects. Because of the small department size and lack of graduate students in most colleges, geology majors have opportunities to assist faculty members in their research projects and in teaching introductory labs.

Universities have programs for graduate study, although not all universities offer graduate degrees in geology; this is something you will want to find out about when you are making your decision. Geology departments in large universities may have hundreds of undergraduate majors and graduate students. In 1980–81, the University of Texas at Austin reported having 582 undergraduate geology majors and 204 graduate students. Obviously, classes in such schools will have many students and sometimes several sections. Undergraduate students will not receive the same amount of personal attention in their geology classes (or in other courses) in a large university that they would in a small college. Introductory classes and laboratories may be taught by graduate students rather than professors. However, the variety of courses offered is likely to be extensive. Large universities often have modern equipment, but access may be restricted to faculty and graduate students. Undergraduate geology majors may have opportunities to work as field or research assistants for graduate students, but probably not for faculty members.

In different ways, small colleges and large universities are both limited by their size in what they can offer students; there is also a continuous range of school and department sizes between these two extremes. Many geology departments, even in large universities, are close-knit groups, more friendly, more open, and more personal than many other departments. You need to consider what will be best for you.

The location of a school can also be an important factor in your decision. Fortunately for earth scientists, geology is everywhere. Whether it involves looking at glacial deposits in Illinois, carbonate rocks in Florida, fault zones in California, or geothermal activity in Wyoming, every part of the country has interesting geology. Financial considerations may keep you near home or in the state where you are a resident; the chance to travel may draw you to a part of the country that you have always wanted to visit. Your college years are a good time to see new places, and choosing a school with a location that interests you may be a way to do this. Many geology courses have field trips to look at "natural laboratories," so you are certain to see and learn about the area around your college. Some schools offer longer field trips for course credit. For example, the University of Vermont visits the Rocky Mountains each year, and Syracuse University has a field trip to a foreign country every spring.

You should also consider your own interests when choosing a college. Some people pick a college because there is a professor there they would like to study with. If you are interested in volcanoes, a department best known for its program in glacial geology might not have a lot to offer you (although you might also discover new talents in glaciology). If you know that you want to learn about petroleum exploration, look for a school that has a strong program in that field. Faculty members in most geology departments can offer suggestions about this.

A further consideration in any choice of college will be finances. As federal aid to higher education is being reduced, the financial burden is falling on students and their parents. Fortunately, geology departments are more likely to be able to offer student aid than some other academic departments. Opportunities may include jobs in the department working with sample collections or drafting illustrations for faculty members, assisting faculty with their research, undergraduate research grants, scholarships from professional organizations, and internships with local companies or government agencies. When requesting information about a school and its earth-science programs, be sure to ask about possible sources of financial aid.

If you already plan to be a geoscientist, you may be tempted to take only science courses in college. Resist this temptation! College is a time and place to obtain a background in many different fields. Most people never have the time or inclination to return to college after they graduate, and their college experience may be the only exposure they will have to a wide variety of ideas. Take the opportunity to explore other fields that interest you besides science: philosophy, music, languages, and so on. Beware of specializing too soon (remember that this is a criticism by some employers of people who have a PhD—"too specialized"). Most colleges will help you get a rounded education by their requirements concerning the "distribution" of courses among the disciplines. An undergraduate program for an earth science major generally involves about 25% of the student's time spent studying his or her special field, about 35% studying related sciences, mathematics, and engineering, and the rest of the time working on general academic subjects.

There are several areas outside of science and mathematics that are especially important for scientists to know about. The most important is English and communications skills. This has been stressed before and will be emphasized again here. It has been rightly said that "The English language is the most important scientific tool we have; use it with precision." In the professional world of earth scientists, the difference between a brilliant idea expressed poorly and a poor idea is irrelevant. No one will listen to either, because no one can understand it. Being concise and clear is an absolute necessity, and you should work hard at achieving that. If composition courses are available, you might try your hand at them. However, your best experience should be within science departments. You may have to write many reports (lab and field) and papers, both on literature research and on your own experiments and field work. You should take the time and effort to do a good job on these. Some graduate schools and employers may even ask to see a sample of your writing, and it is useful to have an example that you're proud of to show them. Gar Williams, of the U.S.G.S., has summed the importance of writing ability to an earth scientist: "Writing can and should be fun, and in today's world it is difficult for a scientist who does not write (and publish) to remain employed."

Another useful subject is economics. Financial considerations are, more and more, determining the directions in which earth science is heading; the decontrol of oil and natural gas prices, fluctuations in the price of gold and the rising cost of silver recovery, and the complexities of natural-hazard insurance and protection are all useful things to understand. Only by understanding the economic significance of the geologist's role in modern life can it be fully appreciated.

Majoring in two subjects in college can make a student more employable on graduation or pave the way for a career shift later. Some students major in two related subjects; for example, a double major in geology and physics would increase a recent graduate's employment prospects in geophysics over a single major in either field. Double majors in unrelated fields can also prove useful. Geology and a foreign language can lead to lucrative overseas assignments with a large consulting or petroleum company. Oceanography and economics might help an earth scientist working in sea-floor mining.

Sigfried Muessig, of Getty Minerals Company, has spoken rather eloquently about the training needed for a career in geology. His remarks, given at the 1980 Geological Society of America meeting, were primarily directed toward planning a career in the mining industry, but his comments are broadly applicable. The person who will have the greatest success, he said, will be a well-rounded, articulate individual with lots of field experience: in other words, a generalist. Specialized knowledge is far less important than the ability to think clearly and logically, using a broad base of geology, supported by mathematics, physics, and chemistry. An example he cited was the wisdom of taking a course in geomorphology, even for someone already planning on being a theoretical igneous petrologist, because geomorphology is an excellent way to learn how to think logically and to understand how scientists arrive at solutions to problems.

Dr. Muessig also stressed the importance of having a professional attitude. This can really only be taught by (and learned from) examples, and thus there is a great need for role models. People who have a professional attitude toward their careers reflect pride in their profession. In displaying a professional attitude, you serve two purposes: you are demonstrating your own feelings about the importance of your chosen career, and you are also providing a role model for other geoscientists, including your peers. An important aspect of professionalism is to see the relationship between a job—a profession—and society as a whole. It is also important to learn and practice the ethical conduct needed in the earth sciences. Being an earth scientist is not just a job, but a role within society. This distinction makes earth science a profession and a career.

Along with this professional attitude, it is important to participate in professional societies. Many of these have special student memberships, which have the distinct advantage of reduced rates for membership dues and discounts on many of their publications. Nearly all have both national and regional meetings, and these are excellent places to meet, talk with, and get to know professional scientists. Even if you are not yet looking for a job, you can plant seeds at these occasions: "I'll be graduating in two years and I am interested in finding out more about your company/office/agency/school/etc." When the time comes and you are sending letters of application and resumes into the job market void, it is extremely helpful if you can address your letter to an individual who will remember your name.

The Graduate School Question

Two factors work in opposition when you graduate from college. On one hand, a master's degree will quickly advance your career in the earth sciences, and there is no better time to enter graduate school than now. All your undergraduate work is fresh in your mind, and your professors are readily available to write letters of recommendation for you. On the other hand, industry is making salary offers to graduating seniors that are very generous, and the idea of earning real money instead of enduring several more years of school may sound very attractive. Which should you choose?

There is no right answer to this question. Many people continue straight into a graduate program and do fine; others are simply too "burned out" from academic work to cope with the additional demands of graduate work right then. Similarly, some people have found that a few years of work experience help clarify their goals and interests so that they can profit much more when they do go back to school. Some alternate working with graduate study. Still others never quite manage to make the effort to return. And there are some earth scientists, of course, who know they never intend to collect an additional degree and are perfectly content with that decision. As with choosing an undergraduate institution, A.G.I.'s Directory of Geoscience Departments can be extremely valuable in selecting a graduate school. In that book, you can find the special interests of the faculty members, the sizes of departments, and addresses for more information.

In making the decision about whether or not to go to graduate school, you need to evaluate your own attitudes and motivation. No one else can make the decision for you. Among the questions that you should ask yourself are:

- How do I feel about two or three more years of school work?
- Knowing how I've acted in the past, what are the chances that, if I do get a job and plan to work for a few years before graduate school, I will actually have the motivation to return to school (and give up the salary)?
- How important is a master's degree or doctorate for what I hope to do in my career?
- How well do I understand my career goals now? Would a few years of work experience help define these better?
- Are there still areas of geology and earth science that I would like to explore before I enter the job market, either because I don't feel well prepared in that area or because I would like to study it in still greater depth?

Field assistant backpacking a mountain precipitation gauge on South Cascade Glacier, Washington. (Photo courtesy of the U.S. Geological Survey.)

The Graduate School Experience

The work involved in graduate study varies among different schools, but the basic elements are the same. A master's degree usually requires two years of schooling beyond the bachelor's degree, although there are exceptions that only require one year . . . or three or more years. In most cases, graduate schools that offer only a master's degree (rather than both a master's and a PhD) will require more time than schools where a master's degree is considered only a stepping stone to a further degree. With the demand for earth scientists increasing, many students are choosing to earn only one degree beyond a bachelor's (if they decide to go on at all). A number of schools allow qualified students to enter a PhD program directly rather than earn a master's degree first. This approach demands considerable maturity on the part of the student.

The first year of graduate school is much the same at most schools; it primarily involves coursework, including both new material that may be related to a thesis topic or that fills course requirements for the graduate program and "remedial" courses to fill any gaps in the student's undergraduate education. During this first year, the student looks for a thesis topic that he or she thinks would be interesting and a faculty advisor who might be interested in supervising the research.

For a master's student, the process of earning a degree generally involves a year of coursework, a summer of field work in the field area chosen for study, and then a second year of coursework, additional research, and writing the thesis. When the thesis is complete, it is usually presented orally before the student's advisory committee, followed by a question and answer session in which the student "defends" his research. A similar presentation may be given to the entire department and other interested individuals. Department traditions vary on the exact procedures, including whether the questions must be confined to the thesis topic alone or they may range over any related (or unrelated) topics in earth science.

This procedure is extended and expanded for doctoral dissertations. The requirements are more rigorous, often including demonstrating some skill in one or more languages and passing a "qualifying exam" to be allowed to continue the academic work. When a student passes the qualifying exam (generally taken at the end of two years), he or she is admitted to candidacy for the PhD and begins dissertation research in earnest. The total amount of time to earn a PhD varies with the student, the research problem, and the university, but most average between four and six years.

Because research for a PhD dissertation is more extensive than for a master's thesis, doctoral students spend a considerable amount of time selecting the research problem. Sometimes the problem will be an addition or a logical extension to research that a faculty member has done. Other times, it will be a topic that the student is interested in and the advisor has consented to help on. For a PhD student, the first summer may be spent helping a faculty member with field work, or doing preliminary work to find a research problem for a dissertation.

If the research problem is not one that a faculty member already has outside funding for, the student may need to find grant money to support the research. Obtaining such financial support is usually possible, especially if the research has some practical application (as in the petroleum industry).

Some students considering graduate school worry about the expense, but this is seldom a problem in the earth sciences. Graduate programs in the earth sciences are generally well funded, and most are able to offer financial support to their graduate students. These offers may include payment of tuition, assistantships in research or teaching, or research duties. Nearly all schools can also help students get low-interest educational loans. Few graduate students have ever made money from their stipends—and most learn to be creative with ground beef and macaroni—but nearly all are able to support themselves while pursuing a degree.

Graduate school is a time for building life-long friendships with fellow students. Everyone is in the same situation—trying to live on their graduate stipends, do research, take classes, prepare for exams, finish a degree, and have a little fun in the process. The support and friendship among the students can go a long way toward making graduate school a good experience. A few professional earth scientists claim that their graduate school days were the best times of their lives!

Never Too Late

Few earth scientists begin their college education knowing that they want to follow a particular subject for a career. Many of the examples cited in this book demonstrate this fact. Some people come from related fields; a physicist may decide to concentrate on geophysics, or a mathematician may change emphasis to meteorology. Others will have planned to study for a career in a very dissimilar area—a surprising number of geology majors began college as potential English majors!

Just as it's never too early to begin planning a career, it's never too late to prepare for a career in earth science. One example is Dick Park, a retired employee of a utility company, who took advantage of a tuition-assistance program for senior citizens. At the age of 62, Dick graduated from college and began a new career in geology with the Bureau of Land Management. Dick commented, "Retirement meant that now I had the time to learn about other fields of interest that I had always wanted to know more about."

Sometimes, people who received degrees in other fields decide to change career paths and return to school for new training. Nancy Evans earned a master's degree in biology, and she has been working in the aerospace program for 12 years. As resident leader of the Viking Orbiter Imaging Team at the Jet Propulsion Laboratory (Pasadena, California), Nancy worked with many scientists, including geologists. "Geology is fascinating," she says. "You can't be around geologists and not get excited about the subject." As a result, Nancy has begun taking courses at a nearby college toward a degree in geology. "Studying glacial and periglacial terrains is still a viable skill," she points out. "There are pipelines to be built and minerals to be found in the far north. If planetary exploration continues, Mars and the satellites of Jupiter and Saturn are all candidates for study because they seem to contain quantities of ice." Even though Nancy became interested in geology 12 years after she finished her college education, she is successfully making a career change to earth science. If it's what you want, it's never too late.

Extracurricular Activities and Summer Experience

Activities outside of the laboratory and classroom can provide valuable experience for later employment. Many colleges require or recommend field camps. These field geology courses are usually taught during the summer, but at least one school in southern California offers a January field geology course. Field

Field work is an important part of geology, and field
experiences begin in freshman classes and continue
throughout college. This photo shows an introductory
historical geology class at Dickinson College studying
Ordovician and Silurian sandstones north of Carlisle,
Pennsylvania. (Photo by the author.)

courses are opportunities to produce a geologic map of one or more small areas while learning basic field techniques. The experience involves working outdoors daily, collecting field data, constructing geologic maps and cross-sections, writing geologic reports, and learning to work with other people as field partners. This last attribute of field camps may be one of the most important parts of doing a field project. Working together in the field requires a special kind of professional and personal relationship, and it is sometimes necessary to work with someone on a professional basis with whom you have no mutual personal interests. The ability to deal with this sort of situation—learning to work well with a variety of types of people, regardless of your individual differences—is important in any type of work.

One advantage to the January field camp is that it leaves the summer open for a job, which can be useful both for experience and income (the ability to take advantage of a January course depends also upon the academic schedule at the college you are attending during that year). Some financial aid is available for students attending summer field camps. The National Association of Geology Teachers, with contributions from several petroleum and mining companies, sponsors the Summer Field Scholarship Program. In the summer of 1980, they offered 120 scholarships to students enrolled in summer field camps.

Even though all geology departments do not require a field camp for graduation, some graduate schools insist that incoming graduate students have attended a field camp. New graduate students at these schools sometimes have to make up the field course during the summer between graduation and beginning graduate school.

Another useful way to learn geology is as a field assistant, helping a graduate student or professional earth scientist who is doing his or her summer research, or writing a senior thesis based on your own field mapping. The cooperative field training program, cosponsored by the U.S. Geological Survey and the National Association of Geology Teachers, provides selected students with summer jobs in the Survey. These students are nominated by the directors of about 120 field camps throughout the United States; the students represent the top one or two students from their camps that year. Fifty to seventy-five of the nominees are hired as summer field assistants with the Survey for the following summer.

These positions are a good foot in the door with the U.S.G.S. for later full-time employment.

Most college and university professors choose their assistants from among their own students, but graduate students are often looking for help. The pay may be minimal—often it's only transportation, food, and a tent to share—but the experience is priceless. Remember, when you consider this type of field experience, to scrutinize the situation for yourself. While the professional or graduate student is evaluating whether you would be an appropriate field assistant, you should be deciding for yourself whether this is someone from whom you can take orders for a summer. This particular arrangement can be very rewarding; not only does this type of experience appeal to prospective employers, but it can also result in a lifelong friendship with the person you are working with.

If you are not interested in spending your summer doing field work, there are other types of internships that are available. Many research labs, universities, government agencies, and private companies add students to their summer staffs to work on laboratory, field, or technical problems. Duties may include drafting maps, making thin sections, and performing routine chemical analyses, but, again, the experience and contacts are valuable. A few programs allow students to perform their own research. For example, NASA's Planetary Geology Undergraduate Research Program allows selected undergraduate students to pursue their own research topics, under the supervision of top scientists at universities, U.S.G.S. offices, and NASA installations around the country. Most intern programs are competitive, and it is wise, if you are interested, to start looking and applying early.

There is, of course, the possibility of regular summer employment as well. Because weather conditions permit outdoor work to be done nearly everywhere in the United States during the summer, many companies expand their operations during the summer months. Mining and drilling companies often need extra people, and engineering firms and government agencies may increase their staffs. These jobs, too, are limited and competitive, and so it helps to start early.

Earth Science: Early and Often

The best preparation for a career in the earth sciences is a familiarity with a broad cross-section of the sciences. Making contacts by meeting professional earth scientists and learning about their jobs and research will help increase your experience and encourage a professional attitude.

The optimum degree for employment and advancement in the earth sciences is a master's degree, but entry-level positions are available to graduates with bachelor's degrees; for research and college teaching, a PhD is usually necessary. At all levels, from undergraduates to doctoral students, a broad viewpoint is important. Keeping interests (and options) open helps in the integration of ideas and information that is so important in the earth sciences.

CHAPTER 9

Women and Minorities in the Earth Sciences

> Why should women who choose to stay home and be wives
> and mothers have to put up with such stuff just so these
> liberated women can prove themselves in a man's field?
> —Geologist's Wife
> (letter in "Ann Landers," 1976)

Women in Earth Science

Several years ago, a woman wrote to Ann Landers complaining that her husband, as a geologist with a major oil company, had to spend a great deal of time with his colleagues, including a woman geologist. The wife wrote, "Of course, I'm not worried about the physical attraction, because most women geologists are so ugly they could go lion hunting with a switch." (Ann Landers made a special point of dissociating herself from this statement, for fear of incurring the wrath of female geologists everywhere!) The days in which this stereotype prevailed are fading into the not-very-distant (but increasingly dim) past.

Of approximately 40,000 geoscientists in the United States, about 5,500 (14%) are women. Almost half (49%) of these women are employed by educational institutions, where new opportunities are limited, and only 18% are employed in industry. However, as superstitions and stereotypes fade, women are entering the earth sciences in increasing numbers. Because overall opportunities are better in industry or government than in academia, women, like men, will have better luck with employers other than educational institutions.

In 1981, women accounted for 24% of all students enrolled in earth science courses in the United States (17% in Canada), and it is not unusual for females to outnumber males in a specific geology classroom. Women accounted for 10% of the PhDs awarded in the physical sciences in 1980, up from 4.5% in 1970.

With the changing role of women throughout society, the role of women in geology has changed also. In 1973, the American Geological Institute started the Women Geoscientists Committee, which acts to support women in the earth sciences and to provide them with both role models and moral support. The goals of the W.G.C. are "to encourage women to enter geoscience fields and to increase professional participation and recognition of women in geoscience." One of their programs is designed to aid young women who are considering a career in earth science when making their career decisions. The committee maintains a speakers bureau, job listings, and a communications network. The Women Geoscientists Committee also publishes a free newsletter (their address is listed in the appendix under the American Geological Institute). The committee is supported by the American Geological Institute and individual voluntary contributions.

In 1978, the Association of Women Geoscientists was founded in the San Francisco Bay area, California. Rather than being an appointed committee, this organization is open to any interested person, with officers elected annually from among the members. Its purposes are "to encourage the participation of women in geosciences, to exchange professional and technical information, and to improve the professional growth and advancement of women in the geosciences." The membership of the A.W.G. is now well into the hundreds and growing, with branch chapters in a number of major cities. It publishes a newsletter, and there are separate membership categories (and dues) for students and professionals. (The address for their national headquarters is listed in the appendix.) Activities within the chapters include field trips, lectures on research topics and career advice, and social functions.

There are numerous agencies and foundations that offer scholarship assistance to women, and some are specifically in-

Dr. Louise Levien is Research Geologist for Exxon Production Research Corporation. Her early interest in chemistry led her into a mineralogy course, and she realized that was the type of chemistry she wanted to pursue. Because of Louise's interests in helping others, she has served on the American Geological Institute's Women Geoscientists Committee. Her advice to young people who plan careers in geology includes these comments:

"I think you really have to go after what you want and try not to fear failure too much. It is an amazing feeling (after the fact) to go past what you thought your limits were and to look back at what you accomplished. Unless you try, you'll never know. Don't take all the advice you're given as gospel. Size up situations for yourself, rather than taking another person's word for it. You have to be in control of your own career. Part of success comes from accomplishment, but a large part is also from recognition. Try to present papers at, or at least attend, professional meetings so that other geologists will know who you are."

Dr. Louise Levien
Research Geologist
Exxon Production and Research
W.G.C. Newsletter, Spring 1980

tended for women pursuing careers in the earth sciences; for example, the June Bacon-Bercey Scholarship in Atmospheric Sciences for Women. (Information about this award should be obtained from the American Geophysical Union, Member Programs Division, whose address is in the appendix; other organizations also offer scholarships).

It is illegal for employers to discriminate on the basis of race, religion, or sex, and just as a woman may not be discriminated *against* when she is applying for a job, she also cannot be discriminated *for*. Because of the equal opportunity laws, a prospec-

The number of women receiving PhD's in the earth sciences has more than doubled in the last decade. Sara Heller, shown here as an undergraduate working on her maps for a field project, received her doctorate in 1980 at West Virginia University. (Photo by the author.)

tive employer may not ask you what your race or sex is, but it may be to your advantage, nonetheless, to let an employer know if you are a female or a member of a minority group. If all your other qualifications are the same as another candidate's, you might receive hiring preference if you are a woman or a member of a minority group (this is an important point: IF all the other qualifications are equal).

Minorities

One of the valuable lessons that most earth scientists learn in the process of doing field work—and working with a wide variety of personality types as field partners—is that each person must be treated and evaluated as an individual. Qualities like being able to work hard without complaining are far more important than sex or ethnic background. As a result, discrimination has not been a major problem in the earth sciences. This may also be related to the global nature of the geologist's perspective, which helps geologists see beyond surface appearances.

The American Geological Institute supports a Minority Participation Program that offers scholarship assistance to U.S. citizens and American-born geoscience majors who are American black, native American, or Hispanic. Each year, this program awards about $50,000 in scholarships, in amounts ranging from $250 to $2,000. These scholarships are available to both undergraduate and graduate students. ("Geoscience" in this case includes geology, geochemistry, geophysics, oceanography, hydrology, meteorology, and space science.) The details for application change slightly each year, and so interested students should contact the American Geological Institute/Minority Participation Program for further details (A.G.I. address in appendix). Other government agencies also offer scholarships to minority students to help them in their education toward careers in geology. The National Science Foundation has provided significant help for both minorities and women in the past, but this function of N.S.F. has been sharply diminished in recent years.

The A.G.I. and its associated societies have worked hard to encourage minority students to enter the geosciences. The percentage rose 13% between academic years 1978 and 1979 and 1979 and 1980, but enrollment for all levels (undergraduate and graduate) is still only 3% of the total number of geoscience students in the United States.

Discrimination is illegal, but there is at least one government agency that does give employment preference to members of a specific minority group: the Bureau of Indian Affairs (Department of Interior) gives preference to applicants who are one-quarter or more Indian. [Further information on this can be obtained from the Chief, Branch of Personnel Services, at the Bureau of Indian Affairs (address in appendix).] With the current interest in economic resources on land that has traditionally been controlled by Indian nations, there is an intense need for earth scientists who are native Americans, especially economic geologists and hydrologists, to study the resources of tribal lands. This need is further accentuated by the fact that there were *no* American Indians employed as geologists in this country in 1980! The Native American Information Center at Bacone College (Oklahoma) studied all major professions to find the representation by American Indians and discovered geology was the only major career field with no Indians on record (although dentistry was second, with only six). In 1980–81, however, 39 students enrolled in U.S. colleges and universities were American Indian or Alaskan natives.

This situation will continue to change as geologists continue to find valuable resources on reservations and tribal lands. A current example of this is in the Black Hills of South Dakota. Ownership of this area has been disputed for a hundred years; the Sioux Indians claim it as historically sacred tribal ground, while several private companies and a government agency claim ownership of various parts. At stake is control of the largest gold mine in the western hemisphere and the bulk of South Dakota's oil, gas, and uranium. The role of a geologist in this dispute is clear; the answers are not.

Opportunities for minority- and women-owned consulting companies are excellent. Federal guidelines requiring demonstrations of minority participation in large, federally funded projects have created a demand for such companies that can provide high-quality services. It is important to remember, however, that the nondiscrimination laws specify that no one can be offered employment in preference to a more qualified applicant; only if the candidates are equally qualified can a minority member or a woman be given preference.

The Outlook

Any distinctions that still exist, setting women and minority groups apart from other earth scientists, can be expected to continue to fade with time. The commitment of the federal government toward equal opportunity employment has encouraged this process.

Women have not yet achieved equality in the employment world. Their employment level and job advancement rate are still lagging behind those of their male colleagues. A clear example of this is in higher education. Women represent 26% of the full-time teaching faculty, but the majority of them have the lowest available rank, while only 1% of the full professors of geology are women. Only 7.2% of tenured faculty members holding a PhD are women. Some of this discrepancy is certainly related to the discouragement of women interested in science that has occurred in the past; if current trends continue, the situation should improve. Major changes in the employment status of women geoscientists, especially in education, will probably be slow in coming. Considerably larger gains are expected in employment for women in industry over the next decade.

As a group, women now employed as physical scientists are paid 75 cents for every dollar that their male counterparts earn. The prospects are much better for women who are entering the labor force now. Starting salaries for female earth scientists are about 99 cents compared to the one dollar that men are offered.

A woman or minority member who is considering a career in the geosciences should make a special effort to meet professionals and other students with similar interests. The general requirements for finding a job will be the same as those for all earth scientists; ultimately, people will be hired only because they are good scientists.

CHAPTER 10

Is Earth Science for You?

Go, my sons, buy stout shoes, climb the mountains, search
. . . the deep recesses of the earth. . . . In this way and in
no other will you arrive at a knowledge of the nature and
properties of things.

—Severinus (7th Century)

Nontraditional Careers

The preceding chapters have described some of the traditional career paths in the earth sciences. There are many other possibilities for innovative combinations of other interests with earth science.

One of these combinations is geoscience and law. Many of the problems confronting the science today present important legal questions. For example, water rights in the southwestern United States are extremely controversial, as are mineral and mining rights in some areas. The resolution of legal problems rising from weather modification will need men and women who understand both law and meteorology. Expertise in geology, oceanography, and law will be important in working out an international treaty on mining the ocean floors.

Talent in public communications, combined with a background in the earth sciences, can lead to a career in presenting science to the public. Television and radio meteorologists, science correspondents for newspapers and magazines, and educational script writers all provide a valuable service in helping the public understand geoscience. Earth science affects people's lives in many ways, and it is an important public function to warn people about tornadoes or floods, explain the scarcity of precious metals, or help them understand the mysteries of earthquakes. Combined studies in some aspect of earth science and communications (English, journalism, film production, or broadcasting, for example) can be useful for this approach.

The need for artists with a knowledge of the earth sciences is limited but real. Technical artists can prepare material for laboratory manuals, textbooks, and professional publications; other artists may find work illustrating books and magazine articles or designing backgrounds for commercial or educational films.

Interest in geoscience and history has led some people into work with museums; historical sites from the days of the California gold rush provide employment for individuals with an understanding of both the geology of the mother lode country and an appreciation for the history of the mid-nineteenth century.

A background in geology and a love of adventure has drawn other people into establishing their own outdoor-guide businesses. Leading river-rafting trips, backpacking tours, and mountain climbing expeditions are some of the ways in which these interests can be combined profitably.

Some people with a particular interest in mineralogy have turned their gem and mineral collecting into small businesses also. Specimens can be sold to other collectors, museums and schools, and the general public. An interest in mineralogy may also lead to further work with gemstones, as a cutter, gemologist, or registered jeweler.

The variety of creative combinations of the earth sciences with other interests is endless: one planetary geologist with a knack for computer graphics has established his own company in Hollywood to provide special effects for science fiction movies; a structural geologist with a pilot's license operates an aerial photography service; a woman who majored in both Spanish and geology in college sells scientific equipment in Spanish-speaking areas of the country. All that is needed is the initiative to try your ideas.

A Continuing Interest

The curiosity that makes a good scientist doesn't end at 5 o'clock on Friday—or at the age of 65. No one has yet proven statistically that "geologists live longer" or have more fun in the process, but the circumstantial evidence is abundant. Professor Andrew Lawson of the University of California (Berkeley) made international news when, at the age of 87, he fathered a baby boy. (Lawson reportedly offered to arm wrestle the reporters who interviewed him—but no one accepted!) Another example is Dr. A. O. Woodford of Pomona College; in 1980, he led some of his former students on a field trip—on his own ninetieth birthday!

One "semi-retired" geologist who has continued to be active in the field is Dr. Mason Hill. During his years as a petroleum and structural geologist, Dr. Hill mapped large areas of California, especially around the San Andreas Fault. Although he is officially retired, he still works as a consulting geologist and lecturer, pursuing his interests in faulting and seismic activity. In 1981, the American Association of Petroleum Geologists awarded him the prestigious Sidney Powers Medal. Dr. Hill describes his introduction to geology as "accidental and fortuitous":

"My introductions to the fields of geology, petroleum geology, and structural geology were all accidental. I doubt that I had any special aptitude for geology—I was never a collector of stamps, bugs, or rocks—but fortuitously I had a professor who led me into the field.

"It just happened that Professor A. O. Woodford advised me to take geology when I was a freshman at Pomona College. During my sophomore year, while taking crystallography and mineralogy from him, I decided to major in geology. (My father, a retail merchant, asked how I could make a living with geology—my answer then was 'I don't know.') Woodford gave me an understanding of the elements of geology and especially how, by whom, and under what conditions the science has developed. He aroused my interest and curiosity about answers to unsolved geologic problems.

"I wasn't financially prepared to attend graduate school

in 1926, but Professor Woodford persuaded me, by arranging for a loan at his bank, to get out of a mining job and into the University of California at Berkeley. After one year there, I took a job in petroleum geology with the Shell Oil Company—mapping in the Transverse Ranges [California]. After 15 months of doing surface geology, I returned to school, obtaining a M.A. from Claremont College for a thesis based on mapping I had done for Shell in the western San Gabriel Mountains. I spent another 15 months with Shell before resuming graduate studies again at the University of Wisconsin. My mentor there was Professor W. W. Mead, who had taught one year at Berkeley—which had happened to coincide with my year there. Fortunately, I was again allowed to use mapping I had done for Shell, of a faulted area west of Santa Barbara, for my PhD dissertation under Mead. This further led me into structural geology, and particularly into faulting. Professor Mead had a great influence on my career by giving me the background and incentive to pursue the study of faults.

"After two depression years, which I spent as a junior college instructor of geology, mathematics, and psychology, and coaching athletics, I returned to Shell from 1934–1937. Thereafter, I advanced with Richfield Oil Corporation and the Atlantic Richfield Company, from district geologist to international exploration manager, until my mandatory retirement in 1969.

"My retirement years have been spent teaching, consulting (mainly for utility companies on seismic problems), and writing papers for publication (most recently, "The San Andreas Fault: History of Concepts" in the Geological Society of America Bulletin, 1980). These activities have continued to give me satisfaction—after I accidentally became a geologist.

"Serendipity has accompanied me all the days of my geological life. I was fortunate to be able to pursue geologic studies and geological society affairs while working as a company petroleum geologist; I am now happy in consulting and retirement geology, and I can't remember any geologic activities or responsibilities that I did not enjoy—geology is fun!

Geologist entering crater of Mount St. Helens, Washington, less than three weeks before its explosive eruption on May 18, 1980, to collect samples. (Photo courtesy of the U.S. Geological Survey.)

"Accidentally, I made a very wise choice of geology as a career. My teachers and associates contributed greatly to my achievements, and, luckily, my strong motivation continues to make the study and application of geology a fulfilling career."

How to Get a Job

Penny Hanshaw, Deputy Chief Geologist with the U.S.G.S. at their headquarters in Reston, Virginia has described how to get a job with her agency. Her comments were specifically directed toward jobs in the U.S. Geological Survey, but they can be applied to getting almost any type of job. Dr. Hanshaw's advice, which was presented at the 1980 national meeting of the Geological Society of America, includes the following points:

- Make contacts before you need to find a job. Professional meetings are an especially good place to meet people, and most professional societies have special categories for student memberships. Your chances of getting a job can be improved greatly if you know people working in the field already, and especially if the people recognize your name when you send in an application.
- Make an appointment for an interview. Don't just drop in.
- Be neat, clean, and well dressed for your interview. Field work sometimes makes earth scientists almost unrecognizable beneath the mud, dust, and rock chips, but most earth scientists make a clear distinction between their "field clothes" and what is presentable in public. Geologists, especially, often dress casually on the job; you should remember that they already *have* jobs. While you are still looking for a position, wearing a suit (whether you are a man or a woman) cannot hurt your job chances. After you accept a position, then you can adapt to whatever the office or laboratory standards may be.
- Study the U.S.G.S. yearbook before your interview to familiarize yourself with the type of projects that are going on within the survey and who is working on them (nearly all companies have annual publications with similar information—write and ask for their most recent annual report). If you have a few intelligent questions about current research projects, or just sound interested in their specific research, you will make a better impression.
- Present evidence of good writing skills. This means that

every piece of writing that you send to a prospective employer should be as well written as possible—from cover letters for your resume to thank you notes after your interview. If you have written a paper or report for another job or class that was particularly well done, you might offer a copy of this as a writing sample for your application file.

- Demonstrate your ability to complete projects on time. Since many projects in the survey or industry are done fairly independently, you will be responsible for many of your own deadlines. The ability to finish projects on schedule is partially a result of being able to budget your time and partly an awareness of the limitations of the circumstances. In nearly all jobs, compromise between the assigned project and limited time means doing the best job possible in the time available. Careful work and attention to detail are important, but a perfectionist may be eternally frustrated.
- Show that you can work well on a team. Nearly all projects with the federal government—and most with industry—involve working with other people, both scientists and technicians. The ability to do this comfortably is a great asset.
- Work to get good grades as an undergraduate. The standard pay scale for federal government work depends upon education and experience, and you can get a higher government service (G.S.) rating if you had good grades (meaning better than a 2.9 grade point average in all your courses and better than a 3.5 grade point average in your major). For more information on where to start looking for available jobs, refer to Appendix D.

Earth Science as a Profession

Why pursue geoscience at all? Isn't a job simply an occupation that keeps you busy during the day and pays you enough to live on? Geology and earth science can be more than that; science is a profession, and as an earth scientist, you have a relationship with—and an effect on—society as a whole.

This dry stream bed in the Sierra Madre Mountains (California) floods during the winter rains. Dr. David Pieri (left), senior scientist at the Jet Propulsion Laboratory, and Dr. Dallas Rhodes of Whittier College are considering evidence of stream activity during those periods. (Photo by the author).

Developing an attitude of professionalism is a slow process, but it is never too early to start. A professional attitude can really only develop through learning by example, and thus there is a tremendous need for role models. If the teachers and scientists whom you know and work with, whether as their student, assistant, or colleague, have ethical and responsible, broad views of their role as a scientist within society, you can hardly help developing a similar attitude yourself. This professional attitude reflects a pride in the profession itself, as well as a sense of its importance to society.

At a time when the earth's finite resources are becoming harder to find and extract, people who can locate the resources and figure out how to use them are becoming national assets themselves. Earth scientists can help protect people from natural hazards, help dispose of hazardous waste, and protect the environment. Think of all the accoutrements of modern society: transportation systems (which run on fossil fuels), buildings (which require metals and rocks), education (chalk, slate, and graphite are common earth materials), even music and the arts (where metals are used in musical instruments and assorted minerals are necessary in paint pigments)—all of these have a fundamental dependence on the geosciences. Thus, the role of the geoscientist is extremely valuable to society.

When you are considering any job, there is a unique set of questions that you should be asking yourself: Will I be doing something that I like? Is it something that I do well? Will I like living in the city or region where the job is? Does the salary meet my expectations?

When you are choosing a career, the questions that you ask should be somewhat different, because a career is more than simply a job. A career is a life-long calling, and it should reflect your personal ideas about your life and what you hope to do with it. The questions that you should consider include: Will my work contribute to society? Can I improve the quality of people's lives? Can I increase man's understanding of the natural world? Will I feel satisfied in this career?

A career in geology or the earth sciences can be as rewarding as you make it. The opportunities for a significant contribution to society exist, and you only need to recognize them.

A number of earth scientists started their careers in science and then moved on to other types of careers where their training permitted them to contribute directly to society. Dr. Harrison Schmitt is a geologist whose interests led him into NASA's Apollo program; he still holds the record for the most remote field area— on the moon! Since leaving the astronaut program, he has served his country in a different way, as a member of the U.S. Senate from New Mexico. Another example is Dr. Frank Press, who was a professor of geophysics; under the Carter Administration, he became the president's science advisor, with a direct influence on federal policies concerning science. Other earth scientists have moved from teaching into high-ranking academic administrative positions. Dr. Frank Rhodes and Dr. James Zumberge, both geologists, are presidents of Cornell University and the University of Southern California, respectively. Despite their other duties and responsibilities, all of these earth scientists maintain an active interest in their original scientific fields.

There are many jobs in the earth sciences, and their availability is expected to increase to 1990 and beyond. However, a career can be much more fulfilling than a job—and the difference between the two is largely in your attitude toward your work. If you have to drag yourself to the office on Monday mornings and watch the slow-moving minute hand until you can leave in the afternoons, perhaps you should look for another type of job—or a different career. But if you find yourself looking forward to work, enjoying your research, liking the people you are working with, and feeling like you're making a positive contribution to society . . . congratulations! You've found yourself a career in earth science!

INTERPRETIVE COMMENT

As suburban growth continues apace, its subtle and delayed effects on the river environment—rate of process, characteristics of channels and flood plains, sediment movement, and aesthetic values—are all going to be affecting more areas and more people. In the small basin discussed here, the picnic places where I took my children are now muddy trash heaps.

Where we played catch, there is now a shrubby and scrubby jungle. The little stream is littered with bricks, concrete trash, plastic bottles, and old tires. Nearby, new and expensive houses look out on a brown mudhole in a small silt-control basin constructed by the builder.

If we are to devise ways in which urban development may proceed with a minimum of these adverse effects, we must have facts—observations made on the ground documenting effects of particular actions. . . . Geologists, more so than most people, know how the natural world operates, and what beauty lies in these mechanisms of nature. If some of the beauty of undisturbed processes is to exist within the reach of cities, the present practices of planning design and construction must include some geologic knowledge. That knowledge can come only from us.

Luna B. Leopold, 1973
"River channel changes with time: An example"
(Address as retiring president of the
   Geological Society of America)
Geological Society of America Bulletin,
   v. 84, p. 1858.

Appendix

Addresses of Geological and Earth-Science Organizations

The following are addresses for professional societies, institutes, and foundations where you can get further information about careers in geology and the earth sciences. Many of these offer brochures or booklets on careers, and most of these are free of charge. Some of these publications are noted in the listing below.

American Association of Petroleum Geologists
Box 979
Tulsa, OK 74101

A booklet on "Careers in Geology" that was produced with the American Geological Institute is available free of charge.

American Geological Institute
One Skyline Place
5205 Leesburg Pike
Falls Church, VA 22041

(Same address for Women Geoscientists Committee and A.G.I.'s Minority Participation Program)
The Women Geoscientists Committee publishes an excellent booklet of career information for all interested students. This material is available free to schools, libraries, and geology departments. It includes information on selected branches of geology and sources of financial aid.

Also available from A.G.I. are a career booklet "Geology: Science and Profession" ($1.00 each), "Careers in Geology" (published with AAPG; free), and the "Directory of Geoscience Departments" ($10.00), which lists over 700 colleges and universities in the U.S. and Canada with courses in the geosciences. Faculty members, their specialties, and field camps are also shown.

American Geophysical Union
2000 Florida Avenue, NW
Washington, DC 20009
A.G.U. publishes two free booklets, "Careers in Geophysics" and "Geophysics: The earth in space."

American Institute of Professional Geologists
Box 957
Golden, CO 80401

American Meteorological Society
45 Beacon Street
Boston, MA 02108

Two career guidance books are available from the A.M.S.: "The Challenge of Meteorology" and "Opportunities in Meteorology" ($5.75).

American Society of Limnology and Oceanography
c/o Department of Zoology
University of Michigan
Ann Arbor, MI 48104

American Society for Oceanography
1730 M Street NW, Suite 412
Washington, DC 20036

The society offers an Oceanography Information Kit for $3.00.

Association of American Geographers
1710 16th Street, NW
Washington, DC 20009

Association of Earth Science Editors
U.S. Geological Survey
Mail Stop 303
Federal Center
Denver, CO 80225

Association of Engineering Geologists
8310 San Fernando Way
Dallas, TX 75218

A.E.G. publishes a free booklet titled "Careers in engineering geology."

Association of Women Geoscientists
Box 1005
Menlo Park, CA 94025

Geochemical Society
c/o David A. Hewitt
Dept. of Geological Sciences
Virginia Polytechnic Institute and State University
Blacksburg, VA 24061

Geological Association of Canada
Dept. of Earth Sciences
University of Waterloo
Waterloo, Ontario N2L 3G1

Geological Society of America
3300 Penrose Place
P.O. Box 9140
Boulder, CO 80301

The booklets that G.S.A. publishes summarizing presentations at their annual meetings on "Future employment opportunities in the geological sciences" are excellent reviews (available on request).

International Oceanographic Foundation
3979 Rickenbacker Causeway
Virginia Key
Miami, FL 33149

The foundation publishes "Training and careers in marine science" ($0.50 per copy).

International Union of Geological Sciences
601 Booth Street, Room 177
Ottawa, Ontario K1A O3B

Mineralogical Society of America
2000 Florida Avenue, NW
Washington, DC 20009

National Association of Geology Teachers
c/o Alan Geyer
Pennsylvania Bureau of Topographic and Geological Survey
Department of Environmental Science
Harrisburg, PA 17120

National Coal Association
1130 17th Street, NW
Washington, DC 20036

The career booklet "The Great American Coal Challenge: What Part Will You Play?" is available free.

National Science Teachers Association
1741 Connecticut Avenue, NW
Washington, DC 20009

"Keys to careers in science and technology" can be obtained for $1.00.

National Speleological Society
1 Cave Avenue
Huntsville, AL 35810

Paleontological Society
Dept. of Geology and Mineralogy
Ohio State University
125 South Oval Mall
Columbus, OH 43210

The booklet "Paleontology in a Nutshell" by Donald M. Fisher is available for $0.10 per copy.

Planetary Society
Box 3599
Pasadena, CA 91103

Seismological Society of America
2620 Telegraph Avenue
Berkeley, CA 94701

Society of Economic Geologists
185 Estes Street
Lakewood, CO 80226

Society of Economic Paleontologists and Mineralogists
Box 4756
Tulsa, OK 74104

Society of Exploration Geophysicists
Box 3098
Tulsa, OK 74101

"Careers in exploration geophysics" can be obtained for $0.25 per copy.

Society of Mining Engineers of AIME
(American Society of Mining, Metallurgical, and Petroleum Engineers)
Caller D
Littleton, CO 80123

Free information includes "Penetrating new frontiers with mineral engineers" and the fact sheet, "Careers for engineers in the minerals industry."

Society of Vertebrate Paleontology
Florida State Museum
University of Florida
Gainesville, FL 32611

Soil Science Society of America
677 South Segoe Road
Madison, WI 53711

Addresses of Federal Agencies

Listed below are the addresses of government agencies that regularly employ geologists and earth scientists. Nearly all have career and employment information that they will send on request.

Agricultural Stabilization and Conservation Service
U.S. Department of Agriculture
Washington, DC 20013

Alaska Outer Continental Shelf Office
Bureau of Land Management
U.S. Department of Interior
Box 1159
Anchorage, AK 99510

Bureau of Indian Affairs
1951 Constitution Avenue, NW
Washington, DC 20245

The fact sheet "Information about employment with the B.I.A." is available free.

Bureau of Labor Statistics
U.S. Department of Labor
1515 Broadway, Suite 3400
New York, NY 10036

The Department of Labor publishes a booklet on "Science and your career."

Bureau of Land Management
Denver Service Center
Federal Center Building 50
Denver, CO 80255

B.L.M. distributes career information called "Opportunities in resource management."

Bureau of Mines
U.S. Department of Interior
2401 E Street, NW
Washington, DC 20241

Bureau of Outdoor Recreation
Dept. of the Interior
Washington, DC 20240

Bureau of Reclamation
U.S. Department of Interior
Denver Federal Center
Building 67
Denver, CO 80225

Fish and Wildlife Service
U.S. Department of Interior
18th and C Streets, NW
Washington, DC 20240

Fact sheets are available on employment opportunities with the U.S. Fish and Wildlife Service.

Forest Service
U.S. Department of Agriculture
Box 2417
Washington, DC 20013

U.S. Geological Survey
U.S. Department of Interior
National Center
12201 Sunrise Valley Drive
Reston, VA 22092

Publications available from the U.S.G.S. include an informational booklet, "United States Geological Survey" (free), "Popular Publications of the U.S. Geological Survey" (free), "Astrogeology: Geological Research in Space" ($0.25), "Marine Geology: Research Beneath the Sea" ($0.25), and "Engineering Geology" ($0.50).

National Aeronautics and Space Administration
400 Maryland Avenue, SW
Washington, DC 20546

NASA Astronaut Candidate Program
Code AHX
NASA Johnson Space Center
Houston, TX 77053

National Oceanic and Atmospheric Administration
U.S. Department of Commerce
11400 Rockville Pike
Rockville, MD 20852

National Park Service
U.S. Department of Interior
Interior Building, Room 2328
18th and C Streets, NW
Washington, DC 20240

The National Park Service publishes a booklet on "Career outlines: National Park Service."

National Weather Service
National Oceanic and Atmospheric Administration
U.S. Department of Commerce
8060 13th Street
Silver Spring, MD 20910

Office of Personnel Management
1900 E Street, NW
Washington, DC 20415

The booklet "Working for the U.S.A." is free on request.

Office of Surface Mining
U.S. Department of Interior
1951 Constitution Avenue, NW
Washington, DC 20240

O.S.M. publishes a free fact sheet about the Office of Surface Mining.

Soil Conservation Service
U.S. Department of Agriculture
Box 2890
Washington, DC 20013

Addresses of State Surveys

Addresses of state geological surveys are listed below. These surveys publish a wide variety of information about the state's natural resources, and they may be able to help you contact earth scientists near where you live.

Geological Survey of Alabama
Drawer O
University, AL 35486

Alaska Division of Geological and Geophysical Surveys
3001 Porcupine Drive
Anchorage, AK 99501

Arizona Bureau of Geology and Mineral Technology
845 North Park Avenue
Tucson, AZ 85719

Arkansas Geological Commission
Vardelle Parham Geology Center
3815 West Roosevelt Road
Little Rock, AR 72204

California Division of Mines and Geology
1416 Ninth Street, Room 1341
Sacramento, CA 95814

Colorado Geological Survey
1313 Sherman Street, Room 715
Denver, CO 80203

Connecticut Geological and Natural History Survey
Room 553, State Office Building
165 Capitol Avenue
Hartford, CT 06115

Delaware Geological Survey
University of Delaware
Newark, DE 19711

Florida Bureau of Geology
903 West Tennessee Street
Tallahassee, FL 32304

Georgia Department of Natural Resources
Earth and Water Division
19 Dr. Martin Luther King Jr. Drive, SW
Atlanta, GA 30334

Hawaii Division of Water and Land Development
Box 373
Honolulu, HI 96809

Idaho Bureau of Mines and Geology
Morrill Hall
Moscow, ID 83843

Illinois State Geological Survey
Natural Resources Building
615 East Peabody Drive
Champaign, IL 61820

Indiana Geological Survey
611 North Walnut Grove
Bloomington, IN 47405

Iowa Geological Survey
123 North Capitol Street
Iowa City, IA 52242

Kansas Geological Survey
1930 Avenue A, Campus West
University of Kansas
Lawrence, KS 66044

Kentucky Geological Survey
311 Breckinridge Hall
University of Kentucky
Lexington, KY 40506

Louisiana Geological Survey
Box G
University Station
Baton Rouge, LA 70803

Maine Geological Survey
Department of Conservation
State House Station 22
Augusta, ME 04330

Maryland Geological Survey
Merryman Hall
Johns Hopkins University
Baltimore, MD 21218

Massachusetts Department of Environmental Quality
Engineering
Division of Waterways
1 Winter Street, 7th Floor
Boston, MA 02108

Michigan DNR—Geological Survey Division
Box 30028
Lansing, MI 48909

Minnesota Geological Survey
1633 Eustis Street
St. Paul, MN 55108

Mississippi Geologic, Economic and Topographic Survey
Box 4915
Jackson, MS 39216

Missouri Division of Geology and Land Survey
Box 250
Rolla, MO 65401

Montana Bureau of Mines and Geology
Montana College of Mineral Science and Technology
Butte, MT 59701

Nebraska Conservation and Survey Division
University of Nebraska
Lincoln, NE 68588

Nevada Bureau of Mines and Geology
University of Nevada
Reno, NV 89557

New Hampshire Department of Resources and Economic
Development
James Hall
University of New Hampshire
Durham, NH 03824

New Jersey Geological Survey
CN-029
Trenton, NJ 08625

New Mexico Bureau of Mines and Mineral Resources
Socorro, NM 87801

New York State Geological Survey
Cultural Education Center
Room 3140
Albany, NY 12230

North Carolina Department of Natural Resources and
Community Development
Geological Survey Section
Box 27687
Raleigh, NC 27611

North Dakota Geological Survey
University Station
Grand Forks, ND 58202

Ohio Division of Geological Survey
Fountain Square, Building B
Columbus, OH 43224

Oklahoma Geological Survey
830 Van Vleet Oval, Room 163
Norman, OK 73019

Oregon Department of Geology and Mineral Industries
1005 State Office Building
1400 SW Fifth Avenue
Portland, OR 97201

Pennsylvania Bureau of Topographic and Geologic Survey
Department of Environmental Resources
Box 2357
Harrisburg, PA 17120

Puerto Rico Service Geologico
Apartado 5887
Puerta de Tierra
San Juan, PR 00906

South Carolina Geological Survey
Harbison Forest Road
Columbia, SC 29210

South Dakota Geological Survey
Science Center
University of South Dakota
Vermillion, SD 57069

Tennessee Division of Geology
G-5 State Office Building
Nashville, TN 37219

Texas Bureau of Economic Geology
University of Texas
Box X
University Station
Austin, TX 78712

Utah Geological and Mineral Survey
606 Black Hawk Way
Salt Lake City, UT 84108

Vermont Geological Survey
Agency of Environmental Conservation
Heritage II Office Building
Montpelier, VT 05602

Virginia Division of Mineral Resources
Box 3667
Charlottesville, VA 22903

Washington Division of Geology and Earth Resources
Olympia, WA 98504

West Virginia Geological and Economic Survey
Box 879
Morgantown, WV 26505

Wisconsin Geological and Natural History Survey
1815 University Avenue
Madison, WI 53706

Wyoming Geological Survey
Box 3008
University Station
University of Wyoming
Laramie, WY 82071

Finding the Jobs

CLASSIFIED ADVERTISEMENTS

Sunday newspapers for nearly all large cities carry advertisements for geologists and geophysicists; jobs in geophysics or the petroleum industry are often well represented, along with some engineering or consulting positions.

Chronicle of Higher Education
This weekly newspaper advertises a variety of academic positions, both teaching and administrative, for American colleges and universities.

Geotimes (a monthly newsmagazine of the earth sciences)
American Geological Institute
5205 Leesburg Pike
Falls Church, VA 22041

The "vacancies" listed in the back of GEOTIMES cover many academic, research, and exploration positions; some state or

local government jobs may also be included. Illustrated advertisements for openings in industry may be scattered throughout
the magazine.

Oil and Gas Journal (weekly magazine for the petroleum industry)
The Petroleum Publishing Company
211 South Cheyenne Ave.
Tulsa, OK 74101

The classified advertisements in this publication concentrate
on positions within the petroleum industry.

EMPLOYMENT SERVICE

The Geological Society of America operates an employment
service that tries to match geologists with open positions. The fee
for a one-year enrollment and employment search is $15, and
applications can be obtained from:

Ms. Joan Heckman
Geological Society of America
3300 Penrose Place
Boulder, CO 80301

DIRECTORIES

Most of these directories to companies are expensive, but
they should be available in large public libraries and universities.

The Geophysical Directory
The Geophysical Directory, Inc.
Box 13318
Houston, TX 77019

The Geothermal World Directory
18014 Sherman Way
Reseda, CA 91335

Industrial Research Laboratories of the United States
R. R. Bowker Company
1180 Avenue of the Americas
New York, NY 10036

Mines Register
American Metal Market Company
576 Fifth Avenue
New York, NY 10026

The Oil and Gas Directory
P.O. Box 13508
Houston, TX 77019

U.S.A. Oil Industry Directory
The Petroleum Publishing Company
211 South Cheyenne Avenue
Tulsa, OK 74101

Addresses of Geology Departments

This section lists some of the four-year colleges and universities in the United States that offer undergraduate (bachelor's) degrees in geology. The list was compiled from two publications of the American Geological Institute: *Directory of Geoscience Departments* (1980–1981) and *Student Enrollment in Geoscience Departments, 1980–1981.* Schools that did not list any geology majors in 1980–81 are not included in this listing.

Further information about the schools and their geology or earth-science programs can be obtained by writing directly to the department. The AGI *Directory* is published annually (AGI's address appears in this appendix under "Directories") and also gives information about the faculty members in each department and their specialties.

ALABAMA

Department of Earth Science
University of Alabama, Birmingham
Birmingham, AL 35294

Department of Geology and Geography
University of Alabama, Tuscaloosa
Tuscaloosa, AL 35486

Department of Geology
Auburn University
Auburn, AL 36830

Department of Geology and Geography
University of South Alabama
Mobile, AL 36688

ALASKA

Division of Geoscience,
    Geology/Geophysics Program
University of Alaska
Fairbanks, AK 99701

ARIZONA

Department of Geology
Arizona State University
Tempe, AZ 85281

Department of Geosciences
University of Arizona
Tucson, AZ 85721

Department of Geology
Northern Arizona University
Flagstaff, AZ 86011

ARKANSAS

Department of Physical Sciences—Geology
Arkansas Tech University
Russellville, AR 72801

Department of Geology
University of Arkansas, Fayetteville
Fayetteville, AR 72701

Department of Earth Science
University of Arkansas, Little Rock
Little Rock, AR 72204

CALIFORNIA

Department of Geological and Planetary Science
California Institute of Technology
Pasadena, CA 91125

Department of Geology
California Lutheran College
Thousand Oaks, CA 91360

Department of Geology
California State University, Sonoma
Rohnert Park, CA 94928

Department of Physics and Earth Sciences
California State College, Bakersfield
Bakersfield, CA 93309

Department of Earth Sciences
California State Polytechnic University
Pomona, CA 91768

Department of Geological and Physical Sciences
California State University, Chico
Chico, CA 95926

Department of Geology
California State University, Fresno
Fresno, CA 93740

Department of Earth Science
California State University, Fullerton
Fullerton, CA 92634

Department of Geological Sciences
California State University, Hayward
Hayward, CA 94542

Department of Geological Sciences
California State University, Long Beach
Long Beach, CA 90840

Department of Geology
California State University, Los Angeles
Los Angeles, CA 90032

Department of Geosciences
California State University, Northridge
Northridge, CA 91330

Department of Geology
California State University, Sacramento
Sacramento, CA 95819

Department of Geology and Geophysics
University of California, Berkeley
Berkeley, CA 94720

Geology Department
University of California, Davis
Davis, CA 95616

Department of Earth and Space Sciences
University of California, Los Angeles
Los Angeles, CA 90024

Department of Earth Sciences
University of California, Riverside
Riverside, CA 92502

Department of Geological Sciences
University of California, Santa Barbara
Goleta, CA 93106

Earth Sciences Board of Studies
University of California, Santa Cruz
Santa Cruz, CA 95064

Department of Geology
Humboldt State University
Arcata, CA 95521

Department of Geology
Occidental College
Los Angeles, CA 90041

Department of Geology and Geography
University of the Pacific,
   College of the Pacific
Stockton, CA 95211

Geology Department
Pomona College
Claremont, CA 91711

Department of Geology
University of Redlands
Redlands, CA 92373

Department of Geological Sciences
San Diego State University
San Diego, CA 92182

Department of Geosciences
San Francisco State University
San Francisco, CA 94132

Department of Geology
San Jose State University
San Jose, CA 95192

Department of Geological Sciences
University of Southern California
Los Angeles, CA 90007

Department of Geology
Stanford University
Stanford, CA 94305

Department of Geology
Whittier College
Whittier, CA 90608

COLORADO

Department of Geology
Adams State College
Alamosa, CO 81102

Geology Department
Colorado College
Colorado Springs, CO 80903

Department of Geology
Colorado School of Mines
Golden, CO 80401

Department of Earth Resources
Colorado State University
Fort Collins, CO 80523

Department of Geological Sciences
University of Colorado
Boulder, CO 80302

Division of Natural and Physical Sciences
University of Colorado at Denver
Denver, CO 80202

Department of Geology
Fort Lewis College
Durango, CO 81301

Department of Environmental Geoscience
Mesa College
Grand Junction, CO 81501

Department of Earth Sciences
University of Northern Colorado
Greeley, CO 80639

Department of Geosciences
University of Southern Colorado
Pueblo, CO 81001

Department of Geology
Western State College of Colorado
Gunnison, CO 81230

CONNECTICUT

Department of Physics and Earth Sciences
Central Connecticut State College
New Britain, CT 06050

Department of Geology and Geophysics
University of Connecticut
Storrs, CT 06268

Earth and Physical Sciences Department
Eastern Connecticut State College
Willimantic, CT 06226

Earth Science Department
Southern Connecticut State College
New Haven, CT 06515

Department of Earth and Environmental Sciences
Wesleyan University
Middletown, CT 06457

Earth, Space and Environmental Sciences Department
Western Connecticut State College
Danbury, CT 06810

Department of Geology and Geophysics
Yale University
New Haven, CT 06520

DELAWARE

Department of Geology
University of Delaware
Newark, DE 19711

DISTRICT OF COLUMBIA

Geosciences Division, Department of
    Earth/Life Sciences
University of the District of Columbia
    Van Ness Campus
Washington, DC 20008

Department of Geology
George Washington University
Washington, DC 20052

Department of Geology and Geography
Howard University
Washington, DC 20059

FLORIDA

Department of Geology
Florida Atlantic University
Boca Raton, FL 33431

Earth Science Program, Department of
    Physical Sciences
Florida International University
Miami, FL 33199

Department of Geology
Florida State University
Tallahassee, FL 32306

Department of Geology
University of Florida
Gainesville, FL 32611

Department of Geology
University of Miami
Coral Gables, FL 33124

Department of Geology
University of South Florida, Tampa
Tampa, FL 33620

GEORGIA

Department of Earth Science
Columbus College
Columbus, GA 31907

Department of Geology
Emory University
Atlanta, GA 30322

School of Geophysical Sciences
Georgia Institute of Technology
Atlanta, GA 30332

Department of Geology
Georgia Southern College
Statesboro, GA 30460

Department of Earth Science
Georgia Southwestern College
Americus, GA 31709

Department of Geology
Georgia State University
Atlanta, GA 30303

Department of Geology
University of Georgia
Athens, GA 30602

Department of Geology
West Georgia College
Carrollton, GA 30117

HAWAII

Department of Geology and Geophysics
University of Hawaii
Honolulu, HI 96822

IDAHO

Department of Geology and Geophysics
Boise State University
Boise, ID 83725

Department of Geology
Idaho State University
Pocatello, ID 83209

Department of Geology
University of Idaho
Moscow, ID 83843

ILLINOIS

Department of Geology
Augustana College
Rock Island, IL 61201

Department of Geological Sciences
Bradley University
Peoria, IL 61625

Department of Geophysical Sciences
University of Chicago
Chicago, IL 60637

Department of Geology and Geography
Eastern Illinois University
Charleston, IL 61920

Department of Geography-Geology
Illinois State University
Normal, IL 61761

Department of Geological Sciences
University of Illinois, Chicago
Chicago, IL 60680

Department of Geology
University of Illinois, Urbana
Urbana, IL 61801

Department of Geology
Knox College
Galesburg, IL 61401

Department of Geology
Monmouth College
Monmouth, IL 61462

Department of Earth Sciences
Northeastern Illinois University
Chicago, IL 60625

Department of Geology
Northern Illinois University
De Kalb, IL 60115

Department of Geological Sciences
Northwestern University
Evanston, IL 60201

Department of Earth Sciences
Principia College
Elsah, IL 62028

Department of Geology
Southern Illinois University, Carbondale
Carbondale, IL 62901

Earth Science, Geography, and Planning
Southern Illinois University, Edwardsville
Edwardsville, IL 62026

Department of Geology
Western Illinois University
Macomb, IL 61455

Department of Geology
Wheaton College
Wheaton, IL 60187

INDIANA

Department of Geography and Geology
Ball State University
Muncie, IN 47306

Department of Earth Sciences
DePauw University
Greencastle, IN 46135

Department of Geology
Earlham College
Richmond, IN 47374

Department of Geology
Hanover College
Hanover, IN 47243

Department of Geography and Geology
Indiana State University
Terre Haute, IN 47809

Department of Geology
Indiana University/Purdue University,
    Indianapolis
Indianapolis, IN 46202

Department of Geology
Indiana University, Bloomington
Bloomington, IN 47401

Department of Earth and Space Sciences
Indiana University/Purdue University,
  Fort Wayne
Fort Wayne, IN 46805

Department of Geosciences
Indiana University Northwest
Gary, IN 46408

Department of Earth Sciences
University of Notre Dame
Notre Dame, IN 46556

Department of Geosciences
Purdue University
West Lafayette, IN 47907

IOWA

Department of Geology
Cornell College
Mt. Vernon, IA 52314

Department of Geography and Geology
Drake University
Des Moines, IA 50311

Department of Earth Sciences
Iowa State University of Science
  and Technology
Ames, IA 50010

Department of Geology
University of Iowa
Iowa City, IA 52242

Department of Earth Science
University of Northern Iowa
Cedar Falls, IA 50613

KANSAS

Department of Earth Sciences
Emporia State University
Emporia, KS 66801

Department of Earth Science
Fort Hays Kansas State College
Hays, KS 67601

Department of Geology
Kansas State University
Manhattan, KS 66506

Department of Geology
University of Kansas
Lawrence, KS 66045

Department of Geology and Geography
Wichita State University
Wichita, KS 67208

KENTUCKY

Department of Geology
Eastern Kentucky University
Richmond, KY 40475

Department of Geology
University of Kentucky
Lexington, KY 40506

Department of Geology
University of Louisville
Louisville, KY 40208

Department of Physical Sciences
Morehead State University
Morehead, KY 40351

Department of Geosciences
Murray State University
Murray, KY 42071

Division of Geology
Northern Kentucky University
Highland Heights, KY 41076

Department of Geography and Geology
Western Kentucky University
Bowling Green, KY 42101

LOUISIANA

Department of Geography and Geology
Centenary College of Louisiana
Shreveport, LA 71104

Department of Geology
Louisiana State University
Baton Rouge, LA 70803

Department of Geosciences
Louisiana Tech University
Ruston, LA 71272

Department of Earth Sciences
University of New Orleans
New Orleans, LA 70122

Department of Earth Science
Nicholls State University
Thibodaux, LA 70301

Department of Geosciences
Northeast Louisiana University
Monroe, LA 71201

Department of Geology
University of Southwestern Louisiana
Lafayette, LA 70504

Department of Earth Sciences
Tulane University
New Orleans, LA 70118

MAINE

Department of Geology
Bates College
Lewiston, ME 04240

Geology Department
Colby College
Waterville, ME 04901

Department of Geology
University of Maine, Farmington
Farmington, ME 04938

Department of Geological Sciences
University of Maine, Orono
Orono, ME 04473

Department of Earth Science,
    Physics and Engineering
University of Southern Maine
Portland, ME 04103

MARYLAND

Department of Earth and Planetary Sciences
Johns Hopkins University
Baltimore, MD 21218

Department of Geology
University of Maryland
College Park, MD 20742

MASSACHUSETTS

Department of Geology
Amherst College
Amherst, MA 01002

Department of Geology and Geophysics
Boston College
Chestnut Hill, MA 02167

Department of Geology
Boston University
Boston, MA 02215

Department of Earth Sciences and Geography
Bridgewater State College
Bridgewater, MA 02324

Department of Geography and Earth Science
Fitchburg State College
Fitchburg, MA 01420

Department of Geological Sciences
Harvard University
Cambridge, MA 02138

Earth Sciences Department
University of Lowell
Lowell, MA 01854

Department of Earth and Planetary Science
Massachusetts Institute of Technology
Cambridge, MA 02139

Department of Geology and Geography
University of Massachusetts
Amherst, MA 01003

Department of Geology and Geography
Mount Holyoke College
South Hadley, MA 01075

Department of Earth Sciences
Northeastern University
Boston, MA 02115

Earth Sciences Department
Salem State College
Salem, MA 01970

Department of Geology
Smith College
Northampton, MA 01063

Department of Geology
Tufts University
Medford, MA 02155

Department of Geology
Wellesley College
Wellesley, MA 02181

Department of Geology
Williams College
Williamstown, MA 01267

Department of Geography and Geology
Worcester State College
Worcester, MA 01602

MICHIGAN

Department of Earth Sciences
Adrian College
Adrian, MI 49221

Department of Geological Sciences
Albion College
Albion, MI 49224

Department of Geology
Central Michigan University
Mount Pleasant, MI 48859

Department of Geography and Geology
Eastern Michigan University
Ypsilanti, MI 48197

Department of Geology
Grand Valley State College
Allendale, MI 49401

Department of Geology
Hope College
Holland, MI 49423

Department of Earth Science,
    Mathematics and Physics
Lake Superior State College
Sault Ste. Marie, MI 49783

Department of Geology
Michigan State University
East Lansing, MI 48824

Department of Geology and
    Geological Engineering
Michigan Technological University
Houghton, MI 49931

Department of Geological Sciences
University of Michigan
Ann Arbor, MI 48109

Geology Department
Wayne State University
Detroit, MI 48202

Department of Geology
Western Michigan University
Kalamazoo, MI 49001

MINNESOTA

Department of Geology and Biology
Bemidji State College
Bemidji, MN 56601

Department of Geology
Carleton College
Northfield, MN 55057

Department of Geology
Gustavus Adolphus College
St. Peter, MN 56082

Department of Geology
MacAlester College
St. Paul, MN 55105

Department of Geology
University of Minnesota, Duluth
Duluth, MN 55812

Department of Geology and Geophysics
University of Minnesota, Minneapolis
Minneapolis, MN 55455

Department of Earth Sciences
St. Cloud State University
St. Cloud, MN 56301

Department of Geology
College of St. Thomas
St. Paul, MN 55105

Department of Geology and Earth Science
Winona State University
Winona, MN 55987

MISSISSIPPI

Department of Geology
Millsaps College
Jackson, MS 39210

Department of Geology and Geography
Mississippi State University
Mississippi State, MS 39762

Department of Geology and
   Geological Engineering
University of Mississippi
University, MS 38677

Department of Geology
University of Southern Mississippi
Hattiesburg, MS 39401

MISSOURI

Department of Earth Science
Central Missouri State University
Warrensburg, MO 64093

Department of Geology
University of Missouri, Columbia
Columbia, MO 65201

Department of Geosciences
University of Missouri, Kansas City
Kansas City, MO 64110

Department of Geology and Geophysics
University of Missouri, Rolla
Rolla, MO 65401

Department of Earth Science
Northwest Missouri State University
Maryville, MO 64468

Department of Earth Science
Southeast Missouri State University
Cape Girardeaux, MO 63701

Department of Geography and Geology
Southwest Missouri State University
Springfield, MO 65802

Department of Earth and Atmospheric Science
St. Louis University
St. Louis, MO 63108

Department of Geology
Stephens College
Columbia, MO 65201

Department of Earth and Planetary Sciences
Washington University
St. Louis, MO 63130

MONTANA

Department of Earth Science
Eastern Montana College
Billings, MT 59101

Department of Geological Engineering
Montana College of Mineral Science
   and Technology
Butte, MT 59701

Department of Earth Sciences
Montana State University
Bozeman, MT 59715

Department of Geology
University of Montana
Missoula, MT 59812

NEBRASKA

Department of Earth Science
Chadron State College
Chadron, NE 69337

Department of Geology
University of Nebraska, Lincoln
Lincoln, NE 68588

NEVADA

Department of Geoscience
University of Nevada, Las Vegas
Las Vegas, NV 89154

Department of Geological Sciences
University of Nevada—
   Mackay School of Mines
Reno, NV 89507

NEW HAMPSHIRE

Department of Earth Sciences
Dartmouth College
Hanover, NH 03755

Department of Earth Sciences
University of New Hampshire
Durham, NH 03824

NEW JERSEY

Department of Earth Sciences
Fairleigh Dickinson University
Madison, NJ 07940

Department of Geosciences
Jersey City State College
Jersey City, NJ 07305

Earth and Planetary Environments
Kean College of New Jersey
Union, NJ 07083

Department of Geological and
   Geophysical Sciences
Princeton University
Princeton, NJ 08540

Department of Geosciences
Rider College
Lawrenceville, NJ 08648

Department of Geological Sciences
Rutgers State University, New Brunswick
New Brunswick, NJ 08903

Geology Department
Rutgers State University, Newark
Newark, NJ 07102

Geology Program
Stockton State College
Pomona, NJ 08240

NEW MEXICO

Department of Geology
Eastern New Mexico University
Portales, NM 88130

Earth Science Division
New Mexico Highlands University
Las Vegas, NM 87701

Department of Geoscience
New Mexico Institute of Mining
   and Technology
Socorro, NW 87801

Department of Earth Sciences
New Mexico State University,
   Las Cruces
Las Cruces, NM 88003

Department of Geology
University of New Mexico
Albuquerque, NM 87131

NEW YORK

Department of Earth Sciences
Adelphi University
Garden City, NY 11530

Department of Geology
Alfred University
Alfred, NY 14802

Department of Geology
Brooklyn College (CUNY)
Brooklyn, NY 11210

Department of Earth and Planetary Science
City College (CUNY)
New York, NY 10031

Department of Geology
Colgate University
Hamilton, NY 13346

Department of Geological Sciences
Columbia University
New York, NY 10027

Department of Geological Sciences
Cornell University
Ithaca, NY 14853

Department of Geology
Hamilton College
Clinton, NY 13323

Department of Geology
Hartwick College
Oneonta, NY 13820

Department of Geoscience
Hobart and William Smith Colleges
Geneva, NY 14456

Department of Geological Sciences
Hofstra University
Hempstead, NY 11550

Department of Geology and Geography
Hunter College (CUNY)
New York, NY 10021

Department of Geology and Geography
Lehman College (CUNY)
Bronx, NY 10468

Department of Geology and Geography
Long Island University, C.W. Post College
Brookville, NY 11548

Department of Earth and Environmental
    Science
Queens College (CUNY)
Flushing, NY 11367

Department of Geology
Rensselaer Polytechnic Institute
Troy, NY 12181

Department of Geological Sciences
University of Rochester
Rochester, NY 14627

Department of Geology
Skidmore College
Saratoga Springs, NY 12866

Geology Department
Southampton College
Southampton, NY 11968

Department of Geology and Geography
St. Lawrence University
Canton, NY 13617

Department of Geological Sciences
State University of New York at Albany
Albany, NY 12222

Department of Geological Sciences
State University of New York
   at Binghamton
Binghamton, NY 13901

Department of Geological Sciences
State University of New York
   at Buffalo
Amherst, NY 14226

Department of Earth and Space Sciences
State University of New York
   at Stony Brook
Stony Brook, NY 11794

Department of Earth Sciences
SUNY, College at Brockport
Brockport, NY 14420

Department of Geoscience, Physics
   and Interdisciplinary Science
SUNY, College at Buffalo
Buffalo, NY 14222

Department of Geology
SUNY, College at Cortland
Cortland, NY 13045

Department of Geology
SUNY, College at Fredonia
Fredonia, NY 14063

Department of Geological Sciences
SUNY, College at Geneseo
Geneseo, NY 14454

Department of Geological Sciences
SUNY, College at New Paltz
New Paltz, NY 12561

Department of Earth Science
SUNY, College at Oneonta
Oneonta, NY 13820

Department of Earth Sciences
SUNY, College at Oswego
Oswego, NY 13126

Department of Earth Sciences
SUNY, College at Plattsburgh
Plattsburgh, NY 12901

Department of Geological Sciences
SUNY, College at Potsdam
Potsdam, NY 13676

Department of Geology
Syracuse University
Syracuse, NY 13210

Department of Geology
Union College
Schenectady, NY 12308

Department of Geology and Geography
Vassar College
Poughkeepsie, NY 12601

Department of Natural Sciences:
    Geology Discipline
York College (CUNY)
Jamaica, NY 11451

NORTH CAROLINA

Department of Geology
Appalachian State University
Boone, NC 28608

Department of Geology
Campbell College
Buies Creek, NC 27506

Department of Geology
Duke University
Durham, NC 27708

Department of Geology
East Carolina University
Greenville, NC 27834

Department of Geosciences
Elizabeth City State University
Elizabeth City, NC 27909

Department of Geology and Earth Science
Guilford College
Greensboro, NC 27410

Department of Geosciences
North Carolina State
Raleigh, NC 27607

Department of Geology
University of North Carolina, Chapel Hill
Chapel Hill, NC 27514

Department of Geography and Earth Sciences
University of North Carolina, Charlotte
Charlotte, NC 28223

Department of Earth Sciences
University of North Carolina, Wilmington
Wilmington, NC 28403

Department of Earth Science
Western Carolina University
Cullowhee, NC 28723

NORTH DAKOTA

Department of Earth Science
Minot State College
Minot, ND 58701

Department of Geology
North Dakota State University
Fargo, ND 58102

Geology Department
University of North Dakota
Grand Forks, ND 58202

OHIO

Department of Geology
University of Akron
Akron, OH 44325

Earth Sciences Department
Antioch College
Yellow Springs, OH 45387

Earth Science Department
Ashland College
Ashland, OH 44805

Department of Geology
Bowling Green State University
Bowling Green, OH 43403

Department of Earth Sciences
Case Western Reserve University
Cleveland, OH 44106

Earth Science Department
Central State University
Wilberforce, OH 45384

Department of Geology
University of Cincinnati
Cincinnati, OH 45221

Department of Geological Sciences
Cleveland State University
Cleveland, OH 44115

Department of Geology
University of Dayton
Dayton, OH 45469

Department of Geology and Geography
Denison University
Granville, OH 43023

Department of Geology
Kent State University
Kent, OH 44242

Department of Geology
Marietta College
Marietta, OH 45750

Department of Geology
Miami University
Oxford, OH 45056

Department of Geology
Mount Union College
Alliance, OH 44601

Department of Geology
Muskingum College
New Concord, OH 43762

Department of Geology
Oberlin College
Oberlin, OH 44074

Department of Geology and Mineralogy
Ohio State University
Columbus, OH 43210

Department of Geology
Ohio University, Athens
Athens, OH 45701

Department of Geology and Geography
Ohio Wesleyan University
Delaware, OH 43015

Department of Geology
University of Toledo
Toledo, OH 43606

Department of Geology
Wittenberg University
Springfield, OH 45501

Department of Geology
College of Wooster
Wooster, OH 44691

Department of Geology
Wright State University
Dayton, OH 45435

Department of Geology
Youngstown State University
Youngstown, OH 44555

OKLAHOMA

Department of Geology
Oklahoma State University
Stillwater, OK 74074

School of Geology and Geophysics
University of Oklahoma
Norman, OK 73019

Department of Earth Sciences
University of Tulsa
Tulsa, OK 74104

OREGON

Department of Geology
Oregon State University
Corvallis, OR 97331

Department of Geology
University of Oregon
Eugene, OR 97403

Department of Earth Sciences
Portland State University
Portland, OR 97207

Department of Geology
Southern Oregon State College
Ashland, OR 97520

PENNSYLVANIA

Department of Geology
Allegheny College
Meadville, PA 16335

Department of Geography and Earth Science
Bloomsburg State College
Bloomsburg, PA 17815

Department of Geology
Bryn Mawr College
Bryn Mawr, PA 19010

Department of Geology and Geography
Bucknell University
Lewisburg, PA 17837

Department of Earth Sciences
California State College of Pennsylvania
California, PA 15419

Department of Geography-Earth Science
Clarion State College
Clarion, PA 16214

Department of Geology
Dickinson College
Carlisle, PA 17013

Department of Earth Sciences
Edinboro State College
Edinboro, PA 16444

Department of Geology
Franklin and Marshall College
Lancaster, PA 17604

Department of Geoscience
Indiana University of Pennsylvania
Indiana, PA 15705

Department of Geology
Juniata College
Huntingdon, PA 16652

Department of Physical Science
Kutztown State College
Kutztown, PA 19530

Department of Geology
La Salle College
Philadelphia, PA 19141

Department of Geology
Lafayette College
Easton, PA 18042

Department of Geological Sciences
Lehigh University
Bethlehem, PA 18015

Department of Geosciences
Lock Haven State College
Lock Haven, PA 17745

Department of Earth Sciences
Millersville State College
Millersville, PA 17551

Department of Geosciences
Pennsylvania State University,
    University Park
University Park, PA 16802

Department of Geology
University of Pennsylvania
Philadelphia, PA 19174

Department of Geology and Planetary Science
University of Pittsburgh
Pittsburgh, PA 15260

Department of Earth and Planetary Science
University of Pittsburgh, Johnstown
Johnstown, PA 15904

Department of Geology
Slippery Rock State College
Slippery Rock, PA 16057

Department of Geological Sciences
Susquehanna University
Selinsgrove, PA 17870

Department of Geology
Temple University
Philadelphia, PA 19122

Department of Geology
Thiel College
Greenville, PA 16125

Department of Geology
Waynesburg College
Waynesburg, PA 15370

Department of Earth Sciences
West Chester State College
West Chester, PA 19380

PUERTO RICO

Department of Geology
University of Puerto Rico
Mayaguez, PR 00708

RHODE ISLAND

Department of Geological Sciences
Brown University
Providence, RI 02912

Department of Geology
University of Rhode Island
Kingston, RI 02881

SOUTH CAROLINA

Department of Geology
College of Charleston
Charleston, SC 29401

Department of Geology and Chemistry
Clemson University
Clemson, SC 29631

Department of Geology
Furman University
Greenville, SC 29613

Department of Geology
University of South Carolina
Columbia, SC 29208

SOUTH DAKOTA

Department of Geology and Geological Engineering
South Dakota School of Mines and Technology
Rapid City, SD 57701

Department of Earth Science and Physics
University of South Dakota
Vermillion, SD 57069

TENNESSEE

Department of Geology
Austin Peay State University
Clarksville, TN 37040

Department of Geography and Geology
East Tennessee State University
Johnson City, TN 37601

Department of Geology
Memphis State University
Memphis, TN 38152

Department of Geography and Geology
Middle Tennessee State University
Murfreesboro, TN 37132

Department of Earth Sciences
Tennessee Tech University
Cookeville, TN 38501

Department of Geosciences
University of Tennessee, Chattanooga
Chattanooga, TN 37401

Department of Geological Sciences
University of Tennessee, Knoxville
Knoxville, TN 37916

Department of Geosciences and Physics
University of Tennessee, Martin
Martin, TN 38238

Geology Department
Vanderbilt University
Nashville, TN 37235

TEXAS

Department of Geology
Baylor University
Waco, TX 76703

Department of Earth Sciences
East Texas State University
Commerce, TX 75428

Department of Geology
Hardin-Simmons University
Abilene, TX 79601

Department of Geology
University of Houston
Houston, TX 77004

Geology Department
Lamar University
Beaumont, TX 77710

Department of Geological Sciences
Midwestern State University
Wichita Falls, TX 76308

Department of Geology
Rice University
Houston, TX 77001

Department of Geological Sciences
Southern Methodist University
Dallas, TX 75275

Department of Geology
St. Marys University
San Antonio, TX 78284

Department of Geology
Stephen F. Austin State University
Nacogdoches, TX 75961

Department of Geology
Sul Ross State University
Alpine, TX 79830

Department of Geography and Geology
Texas A & I University
Kingsville, TX 78363

Department of Geology
Texas A & M University
College Station, TX 77843

Department of Geology
Texas Christian University
Fort Worth, TX 76129

Department of Geosciences
Texas Tech University
Lubbock, TX 79409

Department of Geology
University of Texas, Arlington
Arlington, TX 76019

Department of Geological Sciences
University of Texas, Austin
Austin, TX 78712

Program in Geosciences
University of Texas at Dallas
Richardson, TX 75080

Department of Geological Sciences
University of Texas at El Paso
El Paso, TX 79968

Division of Earth and Physical Sciences,
    Geology Program
University of Texas at San Antonio
San Antonio, TX 78285

Department of Geology
Trinity University
San Antonio, TX 78212

Department of Geosciences
West Texas State University
Canyon, TX 79016

UTAH

Department of Geology
Brigham Young University
Provo, UT 84602

Department of Geology
Utah State University
Logan, UT 84322

Department of Geology and Geophysics
University of Utah
Salt Lake City, UT 84112

Department of Geology and Geography
Weber State College
Ogden, UT 84403

Department of Earth Sciences
Westminster College
Salt Lake City, UT 84105

VERMONT

Department of Geology
Middlebury College
Middlebury, VT 05753

Department of Geology
University of Vermont
Burlington, VT 05401

VIRGINIA

Department of Geology and Geography
James Madison University
Harrisonburg, VA 22807

Department of Geology
Mary Washington College
Fredericksburg, VA 22401

Department of Geophysical Sciences
Old Dominion University
Norfolk, VA 23508

Department of Geology
Radford University
Radford, VA 24142

Department of Geological Sciences
Virginia Polytechnic Institute
   and State University
Blacksburg, VA 24061

Department of Geological Sciences
Virginia State University
Petersburg, VA 23803

Department of Environmental Sciences
University of Virginia
Charlottesville, VA 22903

Department of Geology
Washington and Lee University
Lexington, VA 24450

Department of Geology
College of William and Mary
Williamsburg, VA 23185

WASHINGTON

Department of Geology
Central Washington University
Ellensburg, WA 98926

Department of Geology
Eastern Washington University
Cheyney, WA 99004

Department of Geology
University of Puget Sound
Tacoma, WA 98416

Department of Geology
Washington State University
Pullman, WA 99164

Department of Geological Sciences
University of Washington
Seattle, WA 98195

Department of Geology
Western Washington University
Bellingham, WA 98225

Department of Earth Science
Whitworth College
Spokane, WA 99251

WEST VIRGINIA

Department of Geology
Marshall University
Huntington, WV 25701

Department of Geology and Geography
West Virginia University
Morgantown, WV 26506

WISCONSIN

R.D. Salisbury Department of Geology
Beloit College
Beloit, WI 53511

Department of Geology
Lawrence University
Appleton, WI 54911

Department of Geology
University of Wisconsin, Eau Claire
Eau Claire, WI 54701

Department of Geology and Geophysics
University of Wisconsin, Madison
Madison, WI 53706

Department of Geological Sciences
University of Wisconsin, Milwaukee
Milwaukee, WI 53201

Geology Department
University of Wisconsin, Oshkosh
Oshkosh, WI 54901

Division of Science (Earth Science Program)
University of Wisconsin, Parkside
Kenosha, WI 53141

Department of Geosciences
University of Wisconsin, Platteville
Platteville, WI 53818

Department of Geosciences
University of Wisconsin, Superior
Superior, WI 54880

WYOMING

Department of Geology
University of Wyoming
Laramie, WY 82070

Index

184